Science, Technology and Medicine in Modern History

General Editor: John V. Pickstone, Centre for the History of Science, Technology and Medicine, University of Manchester, England (www.man.ac.uk/CHSTM)

One purpose of historical writing is to illuminate the present. At the start of the third millennium, science, technology and medicine are enormously important, yet their development is little studied.

The reasons for this failure are as obvious as they are regrettable. Education in many countries, not least in Britain, draws deep divisions between the sciences and the humanities. Men and women who have been trained in science have too often been trained away from history, or from any sustained reflection on how societies work. Those educated in historical or social studies have usually learned so little of science that they remain thereafter suspicious, overawed, or both.

Such a diagnosis is by no means novel, nor is it particularly original to suggest that good historical studies of science may be peculiarly important for understanding our present. Indeed this series could be seen as extending research undertaken over the last half-century. But much of that work has treated science, technology and medicine separately; this series aims to draw them together, partly because the three activities have become ever more intertwined. This breadth of focus and the stress on the relationships of knowledge and practice are particularly appropriate in a series which will concentrate on modern history and on industrial societies. Furthermore, while much of the existing historical scholarship is on American topics, this series aims to be international, encouraging studies on European material. The intention is to present science, technology and medicine as aspects of modern culture, analysing their economic, social and political aspects, but not neglecting the expert content which tends to distance them from other aspects of history. The books will investigate the uses and consequences of technical knowledge, and how it was shaped within particular economic, social and political structures.

Such analyses should contribute to discussions of present dilemmas and to assessments of policy. 'Science' no longer appears to us as a triumphant agent of Enlightenment, breaking the shackles of tradition, enabling command over nature. But neither is it to be seen as merely oppressive and dangerous. Judgment requires information and careful analysis, just as intelligent policy-making requires a community of discourse between men and women trained in technical specialities and those who are not.

This series is intended to supply analysis and to stimulate debate. Opinions will vary between authors; we claim only that the books are based on searching historical study of topics which are important, not least because they cut across conventional academic boundaries. They should appeal not just to historians, nor just to scientists, engineers and doctors, but to all who share the view that science, technology and medicine are far too important to be left out of history.

Titles include:

Julie Anderson, Francis Neary and John V. Pickstone
SURGEONS, MANUFACTURERS AND PATIENTS
A Transatlantic History of Total Hip Replacement

Roberta E. Bivins
ACUPUNCTURE, EXPERTISE AND CROSS-CULTURAL MEDICINE

Linda Bryder
WOMEN'S BODIES AND MEDICAL SCIENCE
An Inquiry into Cervical Cancer

Roger Cooter
SURGERY AND SOCIETY IN PEACE AND WAR
Orthopaedics and the Organization of Modern Medicine, 1880–1948

Catherine Cox and Hilary Marland (*editors*)
MIGRATION, HEALTH AND ETHNICITY IN THE MODERN WORLD

Jean-Paul Gaudillière and Ilana Löwy (*editors*)
THE INVISIBLE INDUSTRIALIST
Manufacture and the Construction of Scientific Knowledge

Jean-Paul Gaudillière and Volker Hess (*editors*)
WAYS OF REGULATING DRUGS IN THE 19TH AND 20TH CENTURIES

Christoph Gradmann and Jonathan Simon (*editors*)
EVALUATING AND STANDARDIZING THERAPEUTIC AGENTS, 1890–1950

Sarah G. Mars
THE POLITICS OF ADDICTION
Medical Conflict and Drug Dependence in England since the 1960s

Alex Mold and Virginia Berridge
VOLUNTARY ACTION AND ILLEGAL DRUGS
Health and Society in Britain since the 1960s

Ayesha Nathoo
HEARTS EXPOSED
Transplants and the Media in 1960s Britain

Neil Pemberton and Michael Worboys
MAD DOGS AND ENGLISHMEN (hardback 2007)
Rabies in Britain, 1830–2000

Neil Pemberton and Michael Worboys
RABIES IN BRITAIN (paperback 2012)
Dogs, Disease and Culture, 1830–1900

Cay-Rüdiger Prüll, Andreas-Holger Maehle and Robert Francis Halliwell
A SHORT HISTORY OF THE DRUG RECEPTOR CONCEPT

Thomas Schlich
SURGERY, SCIENCE AND INDUSTRY
A Revolution in Fracture Care, 1950s–1990s

Eve Seguin (*editor*)
INFECTIOUS PROCESSES
Knowledge, Discourse and the Politics of Prions

Crosbie Smith and Jon Agar (*editors*)
MAKING SPACE FOR SCIENCE
Territorial Themes in the Shaping of Knowledge

Stephanie J. Snow
OPERATIONS WITHOUT PAIN
The Practice and Science of Anaesthesia in Victorian Britain

Carsten Timmermann and Julie Anderson (*editors*)
DEVICES AND DESIGNS
Medical Technologies in Historical Perspective

Carsten Timmermann and Elizabeth Toon (*editors*)
CANCER PATIENTS, CANCER PATHWAYS
Historical and Sociological Perspectives

Jonathan Toms
MENTAL HYGIENE AND PSYCHIATRY IN MODERN BRITAIN

Duncan Wilson
TISSUE CULTURE IN SCIENCE AND SOCIETY
The Public Life of a Biological Technique in Twentieth Century Britain

Science, Technology and Medicine in Modern History
Series Standing Order ISBN 978–0–333–71492–8 hardcover
Series Standing Order ISBN 978–0–333–80340–0 paperback
(*outside North America only*)

You can receive future titles in this series as they are published by placing a standing order. Please contact your bookseller or, in case of difficulty, write to us at the address below with your name and address, the title of the series and one of the ISBNs quoted above.

Customer Services Department, Macmillan Distribution Ltd, Houndmills, Basingstoke, Hampshire RG21 6XS, England

Migration, Health and Ethnicity in the Modern World

Edited by

Catherine Cox
Director, Centre for the History of Medicine in Ireland, University College Dublin, Ireland

and

Hilary Marland
Professor of History, Centre for the History of Medicine, University of Warwick, UK

First published 2013 by
PALGRAVE MACMILLAN

Palgrave Macmillan in the UK is an imprint of Macmillan Publishers Limited,
registered in England, company number 785998, of Houndmills, Basingstoke,
Hampshire RG21 6XS.

Palgrave Macmillan in the US is a division of St Martin's Press LLC,
175 Fifth Avenue, New York, NY 10010.

Palgrave Macmillan is the global academic imprint of the above companies
and has companies and representatives throughout the world.

Palgrave® and Macmillan® are registered trademarks in the United States,
the United Kingdom, Europe and other countries.

ISBN 978–1–137–30322–6

This book is printed on paper suitable for recycling and made from fully
managed and sustained forest sources. Logging, pulping and manufacturing
processes are expected to conform to the environmental regulations of the
country of origin.

A catalogue record for this book is available from the British Library.

A catalog record for this book is available from the Library of Congress.

Contents

Tables and Figures

Tables

Figures

Acknowledgements

The articles included in this volume were for the most part developed from papers delivered at a two-day conference held in Dublin on the theme of 'Health, Illness and Ethnicity: Migration, Discrimination and Social Dislocation', which was generously supported by the Wellcome Trust, the Centre for the History of Medicine in Ireland at University College Dublin, Ireland, and the Centre for the History of Medicine at the University of Warwick, UK. We were unable to include all the papers here, but would like to thank all those who attended for providing feedback and lively discussion over the course of the two days which helped shape the contents of this volume. We would also like to take the opportunity to warmly thank the Wellcome Trust for funding our research project, 'Madness, Migration and the Irish in Lancashire, c.1850–1921', which gave us the scope to organise this and other events, as well as supporting the research for our own article. Through their generosity, the Wellcome Trust also supported Roberta Bivins', Kat Foxhall's and Anne Mac Lellan's research projects, the outcomes of which are partly represented by their articles in this volume.

At Palgave Macmillan, Clare Mence and Holly Tyler have offered efficient guidance and excellent support and we would like to thank the series editor for detailed and insightful commentary on each of the chapters.

Catherine Cox and Hilary Marland

Contributors

Alison Bashford is Professor of Modern History at the University of Sydney and elected Vere Harmsworth Professor of Imperial and Naval History at the University of Cambridge, UK. She is the author, most recently, of *Global Population: History, Geopolitics and Life on Earth* (2013) and editor of *Pacific Histories: Ocean, Land, People* (with David Armitage, 2013). She has a long-standing interest in the world history of immigration regulation and its connection with public health. Her 2010 book, co-edited with Philippa Levine, *The Oxford Handbook of the History of Eugenics*, won the Cantemir Prize in 2011.

Roberta Bivins is an associate professor in the history of medicine at the University of Warwick, UK. She is currently completing a monograph on the impact of postcolonial immigration on medical research and health policy in post-war Britain. Her case studies include rickets, smallpox, tuberculosis and the genetic haemoglobinopathies, sickle cell anaemia and thalassaemia. New research explores the twentieth-century domestication of medical technologies from the bathroom scales to the bedside defibrillator. Previous and ongoing work examines the cross-cultural transmission of medical expertise, particularly as exemplified by the transmission of acupuncture to the west (*Acupuncture, Expertise and Cross-Cultural Medicine*, 2000), and in interchanges between medical cultures, both orthodox and heterodox (*Alternative Medicine? A History*, 2007).

Catherine Cox is a lecturer in medical history at University College Dublin and Director of the Centre for the History of Medicine in Ireland, which she co-founded in 2006. She is the author of *Negotiating Insanity in the Southeast of Ireland* (2012), as well as edited collections on medical and Irish history. She has published articles examining institutionalisation in Irish society and on nineteenth-century medical practitioners. Her research interests span medical practice and the medical marketplace in the eighteenth and nineteenth centuries, mental illness in Ireland and England, alternative healing cultures and the relationship between art, disease, the body and medical education in the early nineteenth century. She is currently working with Hilary Marland on a

Wellcome Trust funded project on Irish migration and mental illness in nineteenth-century Lancashire.

Nadav Davidovitch is a public health physician and historian of medicine and public health. He is an associate professor at the Department of Health Systems Management at the Faculty of Health Sciences and Chair of the Center for Health Policy Research in the Negev at Ben-Gurion University of the Negev in Israel. His current research deals with health policy; health inequities; health and immigration; vaccination policy; environmental health and public health history and ethics. Davidovitch serves on several international and national committees, including Chair, Israeli Society for the History and Philosophy of Science; Executive Committee, European Public Health Association; Israel National Advisory Committee for Health Promotion and the Israeli Joint Committee on Environmental Health Policy. He has authored or co-authored over 100 papers and book chapters and co-edited four volumes.

Katherine Foxhall is a lecturer in extra-European History at the University of Leicester, UK, and was a Wellcome Trust Postdoctoral Research Fellow in the Department of History at King's College London, UK. She is the author of *Health, Medicine and the Sea: Australian Voyages c.1815–1860* (2012). Her research has focused on the history of illness experience, quarantine, maritime medicine, emigration and convict transportation. Recent articles have been published in *Social History of Medicine* and *Journal of Imperial and Commonwealth History*, and she is currently working on a history of migraine and a study of quarantine in the Tasman region in the late nineteenth century.

Letizia Gramaglia is an associate fellow in the Yesu Persaud Centre for Caribbean Studies at the University of Warwick, UK. She read English Literature at the Istituto Universitario Orientale in Naples, Italy. In 2001 she was awarded a Rotary Ambassadorial Scholarship and moved to England where she completed her master's degree in Colonial and Postcolonial Literature. She obtained her PhD in Comparative Literature from the University of Warwick, UK, with a thesis focusing on representations of madness in Indo-Caribbean writing. Her research interests and publications focus on colonial and postcolonial literature, women's writing, Indian diaspora and colonial psychiatry, and she produced and wrote an introduction to a new edition of Dr Robert Grieve's *The Asylum Journal (1881–1886)*, which was published in 2010.

Anne Mac Lellan is Director of Research and Academic Affairs at the Rotunda Hospital, Dublin. In 2011, she completed a Wellcome Trust funded PhD in the Centre for the History of Medicine in Ireland, University College Dublin. She is the winner of the 2012 Royal College of Physicians of Ireland History of Medicine Research Award and the joint winner of the 2011 History of Medicine in Ireland essay award sponsored by the Ulster Society for the History of Medicine and the Centre for the History of Medicine in Ireland, University of Ulster. She is a Fellow of the Academy of Medical Laboratory Sciences and contributes a historical column to *Converse*, the Academy's quarterly magazine. She has produced a number of short biographies of Irish women scientists and doctors for two anthologies celebrating the lives and legacies of these women.

Hilary Marland is Professor of History at the University of Warwick, UK, and founding Director of the Centre for the History of Medicine, which she ran from 1998 to 2009. She is the author of *Health and Girlhood, 1874–1920* (2013), *Dangerous Motherhood: Insanity and Childbirth in Victorian Britain* (2004) and *Medicine and Society in Wakefield and Huddersfield, 1780–1870* (1987, 2008), as well as several edited collections on the history of midwifery, alternative medicine, medical practice in early modern England and the Netherlands, maternity and infant welfare, and child health. Her research interests span medical practice and the medical marketplace in the nineteenth century, women and mental illness, alternative healing cultures, the relationship between youth and health, and the history of childbirth and midwifery. She is currently working with Catherine Cox on a Wellcome Trust funded project on Irish migration and mental illness in nineteenth-century Lancashire and is also developing new work on domestic practices of medicine and healing in nineteenth-century Britain, the subject of her next monograph project.

John Welshman was educated at the Universities of York and Oxford and is Senior Lecturer in the History Department, Lancaster University, UK. His research interests are in the history of public policy in twentieth-century Britain, on which he has published widely. Recent articles have been published in *Children & Society* (2008, 2010), *Contemporary British History* (2009), *Economic History Review* (2006), *Journal of Epidemiology and Community Health* (2006, 2007), *Journal of Social Policy* (2004), *Political Quarterly* (2006), *Social History of Medicine* (2006, 2012) and *Twentieth*

Century British History (2005, 2008). His books include *Underclass: A History of the Excluded Since 1880* (2006, 2013), *Churchill's Children: The Evacuee Experience in Wartime Britain* (2010), *Titanic: The Last Night of a Small Town* (2012) and *From Transmitted Deprivation to Social Exclusion: Policy, Poverty, and Parenting* (2007, 2012).

Sarah York is Impact Officer and Teaching Fellow in the History Department at the University of Warwick, UK. She was awarded a PhD in the history of medicine from the University of Birmingham in 2010. She has published articles on the history of psychiatry and suicide and is currently preparing a monograph on the institutional care of suicidal lunatics in nineteenth-century English public asylums. Her present research interests are in the history of military psychiatry and the impact of nineteenth-century warfare on the mental health of British servicemen, as well as narrative medicine and traumatic memory.

Introduction: Migration, Health and Ethnicity in the Modern World

Catherine Cox and Hilary Marland

In recent decades, migration studies has become a vibrant discipline, urged on by the impact of ever more dramatic waves of migration on economies, societies and the provision of services, not least often over-stretched and under-resourced health services. The publication of the latest census returns for England and Wales, for example, showed immigration to be 'larger and greater' than anticipated: by 2011, 56.1 million people lived in the two nations, an increase of 7 per cent over the previous decade, and more than half of the increase was due to immigration.[1] In 2008, fertility rates reached their highest level for 15 years, when figures from the Office of National Statistics revealed that nearly a quarter of babies in England and Wales were born to mothers who came from outside the United Kingdom, particularly women from Pakistan, Poland and India.[2] Other countries have experienced a similar growth in the scale of immigration. The Republic of Ireland, historically an exporter of large numbers of its population, had the highest per capita rate of net migration among European Union member states in 2001. Between 1996 and 2002, approximately 310,700 persons migrated to Ireland representing an 8 per cent population increase; a significant proportion, probably over 46 per cent, was returning Irish.[3] The impact on already struggling health services, especially maternity hospitals and psychiatric support, was substantial and received extensive and often negative press coverage.[4] Meanwhile, the increase in tuberculosis (TB) in Ireland during the last decade has been explicitly linked to the rise in the 'foreign-born' population, leading to calls for the medical screening of immigrants in 2011 – in this instance, 'foreign-born' did not appear to include the families of returning Irish immigrants.[5] A striking feature of the debate has been the insistence that TB has been 'imported' back into Ireland thereby silencing allegations that the poor environmental conditions,

1

previously associated with the spread of the disease, continue to exist in modern Ireland.[6] In Birmingham too, the press reported how TB cases soared to a 30-year high in 2010, equalling rates in 'third world countries'. Tied to 'a major influx of infected people from countries where TB is rife', local councillors and parents called for the reintroduction of TB jabs in local schools.[7] Such scenarios are repeated elsewhere, and not, of course, just in wealthy Western countries, with huge numbers of refugees moving across borders from war-torn countries or to avoid persecution and civil unrest. In many host nations, the impact of migration on health services has been striking and has placed them under a huge strain, with migrants frequently depicted as disadvantaged in terms of health and economic status and thus as a double burden. Yet, through their taxes, young migrants contribute towards supporting aging populations in their new countries, while their labour remains a crucial element in the actual delivery of health care.

A growing body of historical research has engaged in recent years with the relationship between migration and health, as well as the ways in which race and ethnicity impacted on the health of migrants and host communities and on access to health services. Much of this work has focused on migration to the New World, principally North America and Australia, or has examined the medical regulation of borders and disease control on an international scale. Notable among these studies have been Alison Bashford's path-breaking collection of essays, *Medicine at the Border*, which explores 'the pressing issues of border control and infectious disease ... in "the age of universal contagion"', the nineteenth through to the twenty-first centuries,[8] and Amy Fairchild's consideration of the role of medical inspections as both exclusionary and inclusionary tools in the political economy of industrial America.[9] A recurring theme in the literature, particularly in studies of North American immigration policy, has been the tension between the need for immigrant labour and the risk of admitting migrants who would become long-term economic dependants. As Fairchild has argued, 'the overwhelming impulse was to absorb the immigrants into the laboring body'. When '[i]mmigrants were rejected', it was 'mainly for causes related to economic "dependency"'; many of those refused were considered to have existing or potential health problems that limited their labouring capacity and made them likely to become public charges, reliant on state welfare. For immigrants, the potent image of exclusion was the station at Ellis Island where thousands received 'snap diagnosis', carried out by the medical officers of the Public Health Services in line examinations.[10] These officers, however, became part of the process of building a 'highly mobile unskilled labor force'; and indeed only a small proportion of immigrants, less than 1 per cent, were deported on medical grounds.[11]

Rather than taking disease control, border security and the politics of global health as its main focus, this volume explores migrant and host country experiences of immigration and localised responses to the importation of diseases and other health problems. It builds on the collection of essays, *Migrants, Minorities and Health*, edited by Lara Marks and Michael Worboys in 1997, which focused on the ways in which nineteenth- and twentieth-century medicine and medical services responded to immigrants and ethnic minorities in a range of host countries.[12] As in Marks and Worboys' volume, our contributors move away from the strong emphasis in the historiography on North America to examine finely grained case studies taken chiefly, though not exclusively, from the Anglophone world, including Britain, Australia, Israel and the Caribbean. Focusing on the relationship between migration, health and illness over an extended timeframe, c.1820 to the present day, the volume assesses changes in the health status of migrant groups in a period encompassing Imperial expansion, decolonisation and new waves of economic and political migration in the late twentieth century. The chapters emphasise the extent to which policies were shaped and implemented in response to specific health issues linked to specific migrant groups, be these resolution of the challenges of rickets and TB, as outlined by Roberta Bivins, or the irresolute response to the threat of TB among Irish nurses in Anne Mac Lellan's chapter.[13] The timeframe covered by the book enables exploration of the transition from public health concerns, through anxieties about racial decline shaped by fears of degeneration and eugenic discourse, to the impact of biomedicine and biopolitics and the globalisation of health policies. More unusually, it also focuses on chronic illness and the management of mental health and mental incapacity on a global and local level.[14]

Migration is a complex process and no less so when considered in relation to health problems and practices. There are many different forms of migration and types of migrant, and the chapters in this volume consider the diversity of migration prompted or forced (in the examples drawn on by Foxhall and Letizia Gramaglia) by political or economic expediency as well as more positive experiences of migration in response to perceived opportunities. It explores the ways in which ideas on health, and notably health disadvantage, among migrant communities interacted with changing ideas and ideologies of ethnicity and race, and changes within health services themselves, as well as the role of place, policy and individual actors on migrant health experiences. Oftentimes, as demonstrated by the asylum superintendents cited by Gramaglia and Catherine Cox, Hilary Marland and Sarah York, or the key policy makers referred to by Bivins and John Welshman, individuals became influential players in determining both local initiatives and national policy. Debates on migrant health and the implementation of policy in some cases contributed, as noted by Alison Bashford and Bivins, to broader discussions on health and health policy and medical science; in other examples, notably Welshman's analysis of key turning points in debates on social deprivation, they could loom small.

One of the dominant assumptions in connection with migration and health, and one pointed out by Marks and Worboys in the introduction to their volume, is that immigrants are always at a health disadvantage, as many are economic migrants or political refugees, experiencing physical and psychological dislocation while starting in badly paid jobs and poor housing in their new countries, often isolated and subject to language difficulties and discrimination. They are perceived as facing the 'double jeopardy' of being a minority and sick, and also had difficulties accessing services.[15] But, as several of the essays in Marks' and Worboys' volume show, this was not always true; some immigrants might well have health advantages.[16] The chapters in our volume, however, somewhat in contradiction to Marks and Worboys' conclusions, point predominantly to a strong association between migration and health disadvantage. Many migrants were disadvantaged at both the point of departure and arrival, such as the Irish migrants featured in the account of Cox, Marland and York, Gramaglia's indentured labourers or migrants to the new state of Israel post-1948 in Nadav Davidovitch's chapter. Bivins, however, observes that commentators on migrant health appear to have paid scant attention to the issue of poverty in debates on TB and rickets, focusing rather on issues related to ethnicity and cultural practices.

Migrants do not necessarily accept and adopt the dominant medical culture; many retained strong links with their country of origin and brought deeply embedded ideas and practices of health to their host countries, as demonstrated in Davidovitch's and Mac Lellan's chapter; whether they were able to act on these different approaches to health was another matter. Davidovitch's chapter in particular highlights wide divergences in ideas about the utility of vaccination between migrants and the new state of Israel between 1948 and 1956 and also the degree of compulsion employed in enforcing compliance.[17] Variations within groups of migrants, particularly with regard to identity, are problematic too when exploring migrant populations. Migrants are not just migrants, but also have other identities and social positions, based on class, gender, age, occupation and other variables, and country of origin.[18] The case studies in this volume strive to disaggregate specific sets of migrants to throw light on the diversity of experience in terms of ethnicity, social class, gender and age, ranging from the enormous diversity of migrants to Israel in the post Second World War years (Davidovitch) to Indian labourers in British Guiana (Gramaglia), convicts and emigrants voyaging to Upper Canada and Australia (Foxhall), and Irish nurses working in the mid-twentieth-century British health service (Mac Lellan).

Reflecting Fairchild's findings, the requirements of labour markets emerge as a dominant theme in several chapters, particularly when these needs are pitted against anxieties about the risks of migration from the perspective of host communities. Mac Lellan illuminates how dependence on Irish nursing staff in English hospitals, between the 1930s and 1960s, overrode demands to screen and survey for TB in spite of the threat to the health of the majority community and the nurses themselves. The demands of the Lancashire labour market, explored by Cox, Marland and York, drew Irish migrants into the region throughout the nineteenth century; yet large numbers were to become long-term residents in local lunatic asylums further burdening welfare provision and its cost in the county. Likewise, during the years of slave emancipation in Barbados, as Foxhall shows, there were high levels of mobility as plantation owners and labouring populations negotiated the changing labour market. In this new landscape the risk of smallpox outbreaks loomed as quarantine laws were apparently contravened and vaccine supplies were found to be inadequate. In each of these case studies pressures from the labour market dominated and in some instances trumped anxieties about the health of the migrants, the threat to the host community and the potential drain on resources.

The most obvious example of challenges to the health of the host community wrought by migration, particularly mass migration over short periods as in the case of the Famine Irish or migration to post-war Israel – in terms of public perception, government responses and actual experience – is in the domain of public health, notably the risk of epidemic disease.[19] Davidovitch's chapter examines the robust, sometimes heavy handed, approaches adopted by the Israeli state to the diversification of immigration in the post-war era, when concerns about the risk of infection became strongly associated with the ethnicity of post-Holocaust migrants, and notable for the 'racial image' of the newcomers who were perceived as likely to be suffering from diseases such as TB or ringworm.[20] Yet in Foxhall's account certain kinds of migrants – notably children – were regarded as potential agents of public health and tools of medical endeavour; in this case, carriers of live vaccine from countries of departure to the new colonies. In the latter decades of the nineteenth century, anxieties about disease causation were marked by a general shift in perception which increasingly targeted individual lapses in hygienic responsibility and temperate behaviour rather than the broader environment and risks represented by the migration of large groups, often poor and perceived as already diseased and contagious. Thus there was a re-labelling of the idea of 'the other', and this shift was responded to, for example, in Ellis Island controls and medical checks leading to concern with migrants as individual carriers of germs rather than groups collectively producing insanitary conditions leading to disease.[21] These concerns could also be fused or blended together as in the case of late nineteenth-century Irish migrants in Lancashire whose hygienic practices were monitored by public health officials who noted with alarm the tendency of the Irish to inhabit the worst and most insanitary districts in towns and cities and also to put their health at risk through their cultural practices. By the twentieth century, as Bivins and Mac Lellan show for Britain, the Irish continued to be associated with contagious diseases, especially TB, but policy makers generally believed that the problem could be tackled through routine health programmes and did not require specific, targeted campaigns. In contrast, the incidence of TB and rickets among South Asian immigrants was explicitly associated with their cultural practices and they were identified as importers of 'old' contagious and chronic diseases into Britain. In Bivins' case studies, emigration and immigrants were portrayed as inhibiting progress in the development of a modern British state. Health campaigns and policy makers identified Asian cultural habits as contributing to high levels of rickets, maintaining that greater acculturation would act as a prophylactic against the disease.

Such attitudes produced a variety of responses among migrants. Shah has demonstrated the ways in which twentieth-century Chinese-American activists sought to assimilate American norms into domestic arrangements, consumption patterns and social conduct, which went on to produce – as segments of Chinese America assimilated – recognition of citizenship and disbursal of government resources and services to maintain them in good health, though many other Chinese were left behind and noted to be 'aberrant'.[22] Bivins' chapter demonstrates how members of the South Asian community absorbed and accepted as 'reasonable' the associations made between TB and migrants. In contrast, Davidovitch's study highlights cases of resistance among migrants in the new state of Israel in spite of their extremely precarious state of health and the significant resources invested in vaccination campaigns. The story of the relationship between disease and migration can be highly complex as in Anne Mac Lellan's study of tubercular Irish nurses in England between1930 and 1960, which examines how far the role of environment and poor health status, both at the point of departure and at their destination, affected migrant nurses and shaped debates on their liability to illness, showing the fluidity of concepts around environment and individual behaviour. Though the nurses greatly feared TB, they were reluctant to give up their posts, the rewards of their employment finely balanced against the dread not only of contracting a deadly disease but also of the stigma attached to it.

A wide variety of agencies – medical professionals, welfare organisations, government and policy makers, as well as the media and broader publics – concerned themselves with issues related to migration, health and welfare. Several chapters assess the role that medicine played in shaping policy as well as political and public discourse on the impact of migration on host communities and their health policies, practices and outcomes. Notably Bashford's chapter emphasises the role of discourses and practices with regard to insanity and feeble-mindedness as a spur for the enactment of border controls and the pursuance of eugenic policies, a process also outlined in Dowbiggin's study of the responses of US and Canadian psychiatrists to immigration in the early decades of the twentieth century.[23] As Foxhall and Bivins' chapters demonstrate, the movement of large numbers of migrants also provided medical professionals with new opportunities. Foxhall argues that British and Irish emigrants and convicts onboard ships became subjects of naval surgeons' early experiments with vaccination and provided a method of transporting 'live' vaccine matter between the metropole and the colonies. The 're-emergence' of rickets in post–Second World War Britain among the immigrant and second-generation South Asian communities presented a group of elite researchers with a new and apparently compliant pool of 'clinical material' to better understand the causes of the disease which had remained ill-defined.

The volume places a strong emphasis on the impact of mental illness among migrants and host communities, situating this not as discrete discourses and sets of interventions, but rather alongside discussions of other health issues effecting migrant communities and individuals. High rates of mental illness among migrant groups form a continuum across geographical space and time, and, as Bashford demonstrates, played a lead part in shaping international eugenic practice, notably with regard to border controls. Fear of inferior immigrant stock and of the consequences of miscegenation, and the impact of both on national efficiency as well as on often strained budgets and services, fashioned concerns about admitting the mentally unfit over a long time period and in many different contexts.[24] As Bashford demonstrates, from the nineteenth century onwards, a key aspect of immigration restriction laws – and one that has often been overlooked – was the power to exclude or deport individuals on the grounds of insanity. Her chapter assesses these clauses in Anglophone immigration restriction acts in a variety of contexts, including North America, the Australian colonies, New Zealand and the Cape, arguing that it was the standard insanity clause in almost all immigration acts that became a key manifestation of international eugenic practice. For example, in New Zealand the 1866 Aliens Act and the 1873 Imbecile Passengers Act excluded immigrants not of 'sound mind'. Her chapter also examines just how this was, and sometimes was not, also about ethnic and racial exclusion. Meanwhile, taking specific case studies has enabled us to look at the management of mental illness as part of a wider set of concerns with the health of migrant populations, as in Cox, Marland and York's case study of Irish migration into nineteenth-century Lancashire. Here the management of the insane Irish was seen as part of a much larger set of health and welfare issues associated with the 'Irish problem', and a very significant addition to public health anxieties involving the Irish. Gramaglia's essay demonstrates how the old world system of moral management was transferred to the colonies of the Caribbean where reformed asylums were established several decades after their widespread development in the United Kingdom. In an apparently 'enlightened' approach to understanding the stress of migration and the aetiology of mental breakdown, Gramaglia demonstrates how one asylum doctor based in British Guiana towards the end of the nineteenth century attributed high rates of mental illness among migrant workers directly to the dislocation of migration, a factor also pointed out by several asylum superintendents in Victorian Lancashire. Gramaglia's and Cox, Marland and York's chapters alert us to the fact that asylum doctors could be keen observers of migration and its impact on mental health, though – as with Bashford's doctors – they were reluctant to abandon anxieties about the susceptibility of particular 'racial' types to mental disease, as well as addiction to drugs and alcohol.

While Marks and Worboys have suggested that migrants might not necessarily face impediments with regard to health status or access to services and support, the case studies in this volume illustrate strong associations between migration and poor health outcomes. Migrants were described as susceptible to both mental and physical breakdown and disease, with the act of migrating and the risks associated with travel and transience acting as contributory factors, alongside discrimination and social deprivation. Cox, Marland and York's study suggests that the reality of extreme poverty, deprivation and social isolation among Irish asylum patients in Lancashire contributed to their accumulation as chronic long-stay patients. Foxhall's case studies likewise examine groups of migrants who were, due to their extreme poverty and deprivation, at a health disadvantage onboard ships and vulnerable to disease. The 'act' of migrating left them susceptible to illness and, added to this, they became the subjects of medical innovation and diplomatic manoeuvring. While the Irish nurses studied in Mc Lellan's chapter were less vulnerable to the direct impact of migration and less obviously deprived, they found themselves isolated and a long way from home, and their susceptibility to TB was tied to poor working environments compared with those of their English counterparts. In this case, a different form of health 'disadvantage' – limited exposure to disease prior to migration and an underdeveloped vaccination programme in Ireland – increased their vulnerability to TB. Gramaglia's labour migrants faced the combination of disruption and cultural dislocation of migrating and exposure to dire labour conditions, which wrecked both their physical and mental health.

Studies of the relationship between migration and disease has often coalesced around outbreaks of epidemic diseases, illuminating, as in Davidovitch's study of smallpox vaccination, the perceptions of politicians and state officials that the cultural practices of 'primitive' populations allowed the importation of 'old' diseases. Many of the chapters in this volume also draw attention to chronic, slow-burning, less dramatic, endemic diseases – mental illness, rickets and TB. Such disorders shifted policy makers' attention to diseases believed to be entrenched in incoming populations and exacerbated by cultural behaviour. The campaigns devised to eradicate both chronic and epidemic diseases, as Bivins, Davidovitch and Foxhall argue, were often connected to issues of migrant citizenship in their new host country. While Shah has demonstrated how public health reform was far more than an instrument to suppress epidemic disease and enhance human vitality, promoting strategies of governance and citizenship, our case studies demonstrate that these strategies could be coercive and narrowly delimited.[25] The medical professionals involved in vaccination programmes in the new

Israeli state, discussed by Davidovitch, were not only tackling disease, they were creating new, modern citizens by allowing or, as Davidovitch argues, in some cases forcing people to think about their bodies and surroundings in different ways: they endeavoured to change individual behaviour and as Warwick Anderson has suggested gave 'new meanings to mundane interactions with others and the environment'.[26] Yet, Bivins' chapter identifies the limits placed on efforts to include migrants as citizens through health programmes; the South Asian ethnic communities in her study were, to a certain extent, excluded from concepts of 'public' health. In contrast both Bivins' and MacLellan's case studies demonstrate that Irish migrants, though still associated with disease and unhealthy cultural habits, were recognised as part of the British workforce.

Access to health services such as clinics, according to Anderson, helped migrants imagine what a future as citizens could mean for themselves and their families. Fairchild, meanwhile, has argued that while public health officials (and, for Anderson, laboratories) were engaged in disease diagnosis, treatment and prevention, they also participated in shaping the national body by playing a pivotal role in selecting and rejecting future citizens. Clinics and, in Davidovitch's case, the transit camps were also sites where the national body was checked, registered and restored. In Davidovitch's 'mothers and children' clinics Jewish immigrants, often emanating from Asia and Africa during the 1950s, were vaccinated and trained in the hygienic practices of the new Israeli state and participation in vaccination programmes became part of the governance of Eastern-European white and non-white Asia and African immigrants. The health practices of immigrants conceptualised as 'dirty' and 'primitive' were to be substituted by modern forms of health care in the new state of Israel. Doctors, nurses, scientists and public health officers were conceived of as experts in bodily reform and hygiene, white citizenship and national destiny. Yet in most of our case studies, migrants accessed or found themselves in services that were old-fashioned and almost obsolete in terms of effective treatment. In addition, ideas of 'whiteness' were not fixed; Fairchild's study has suggested that whiteness could play less of a role and in the United States 'while...medical and lay immigration officials saw the world in terms of a multitude of unequal races, they placed little emphasis on drawing fine, racial distinctions between peoples with the goal of exclusion'. Rather the 'overwhelming impulse was to absorb immigrants into the laboring body'.[27]

This volume deals less with race than ethnicity, though, building on Shah and Anderson's conclusions, it begins with the idea of whiteness as the 'norm' or the majority group.[28] Bivins and Welshman, in particular, examine the reluctance of governments and social scientists to utilise the language of race in discussing susceptibility to particular forms of illness and outcomes and to poverty, though as Bivins demonstrates race and ethnicity continued to be invoked in discussions on the relationship between illness, cultural practices, environment, diet and behaviour. The chapters also question how long groups or individuals new to a country remain migrants and at what point they segue into becoming members of minority communities. As Dr Rogers, Medical Superintendent of Rainhill Asylum, sagely commented, classifying admissions on the basis of place of birth masked the fact that many patients were ethnically Irish, 'essentially Irish in everything but their accidental birthplace'.[29] Yet Welshman demonstrates that the importance of questions of ethnicity varied in the numerous post Second World War studies of social deprivation in the United Kingdom, while Foxhall suggests that the race of the smallpox 'carriers', a factor that would emerge so strongly in subsequent anti-vaccination campaigns, mattered less in the early nineteenth century.

The chapters thus highlight the diversity of responses to migrants' race and ethnicity that emanated from policy makers, medical professionals and the media in debates about disease susceptibility and access to health services during the two centuries examined in this volume. While it can be argued that concerns about racial degeneration became especially acute at key moments – most notably in the late nineteenth century when anxieties about degeneration peaked – immigrants' experiences of ill health and encounters with health professionals and medical services were not always dominated by these issues. Far from it, as poverty, social isolation, gender, the act of migrating itself and the aspirations of state officials when promoting the requirements of the host country – including the need for labour – could dampen or override anxieties about ethnicity and race. And while Bashford points to levels of unity in developing policies in the specific case of mental health and disability, in many other cases locale proved vital in shaping particular responses to the health challenges and needs of migrant populations.

Notes

1. http://www.bbc.co.uk/news/uk-politics-18869955.
2. http://www.guardian.co.uk/world/2009/aug/27/population-growth-uk-birth-rate-immigration.
3. E. O'Sullivan, *Migration and Housing in Ireland. Report to the European Observatory on Homelessness* (Brussels: Feantas, 2002).

4. 'Seeking Asylum in the Labour Ward', *Irish Independent*, 24 June 2000.
5. D. Pringle, 'The Resurgence of Tuberculosis in the Republic of Ireland: Perceptions and Reality', *Social Science and Medicine*, 68 (2009), 620–4; N. Cahilll 'TB Screening for Immigrants Proposed', *Irish Medical News*, 23 November 2011 [accessed 23 August 2012].
6. But see Sarah Curtis who argues that poor living conditions accounted for the high incidence of TB among ethnic minorities in the United States and Britain: S. Curtis, *Health and Inequality: Geographical Perspectives* (London: Sage, 2004).
7. http://www.birminghammail.net/news/top-stories/2010/12/14/immigration-blamed-for-soaring-tb-in-birmingham-97319-27818787/.
8. Alison Bashford (ed.), *Medicine at the Border: Disease, Globalization and Security, 1850 to the Present* (Houndmills: Palgrave Macmillan, 2006).
9. Amy Fairchild, *Science at the Borders: Immigrant Medical Inspection and the Shaping of the Modern Industrial Labor Force* (Baltimore, MD and London: Johns Hopkins University Press, 2003).
10. Alan M. Kraut, *Silent Travelers: Germs, Genes, and the 'Immigrant Menace'* (New York: Basic Books, 1994); Amy L. Fairchild, 'The Rise and Fall of the Medical Gaze: The Political Economy of Immigrant Medical Inspection in Modern America', *Science in Context*, 19 (2006), 337–56.
11. Fairchild, 'The Rise and Fall of the Medical Gaze', 338; Fairchild, *Science at the Borders*, 4, 10.
12. Lara Marks and Michael Worboys (eds), *Migrants, Minorities and Health: Historical and Contemporary Studies* (London and New York: Routledge, 1997).
13. See also Roberta Bivins, '"The English Disease" or "Asian Rickets"? Medical Responses to Postcolonial Immigration', *Bulletin of the History of Medicine*, 81 (2007), 533–68; Krista Maglen, 'Importing Trachoma: The Introduction into Britain of American Ideas of an Immigrant Disease, 1892–1906', *Immigrants & Minorities*, 23 (2005), 80–99.
14. See also the essays in Angela McCarthy and Catharine Coleborne (eds), *Migration, Ethnicity, and Mental Health: International Perspectives, 1840–2010* (New York and London: Routledge, 2012), which explore the relationship between immigration and mental health and illness over a broad timeframe and a number of geographic arenas.
15. Lara Marks and Michael Worboys, 'Introduction', in Marks and Worboys (eds), *Migrants, Minorities and Health*, 11–12, 6.
16. As Marks and Hilder show in terms of the low rates of infant mortality among Jewish East European immigrants at the turn of the century and Bengali immigrants in the late twentieth century, which they relate to community and familial support as well as cultural factors: Lara Marks and Lisa Hilder, 'Ethnic Advantage: Infant Survival among Jewish and Bengali Immigrants in East London, 1870–1990', in Marks and Worboys (eds), *Migrants, Minorities and Health*, 179–209, while Powles' survey of Greek immigrants to Australia demonstrates that they had better health status than both the broader host population and communities left behind in Greece: John Powles, 'Greek Migrants in Australia: Surviving Well and Helping their Hosts', in idem, 210–27.

17. Though see, Nayan Shah, *Contagious Divides: Epidemics and Race in San Francisco's Chinatown* (Berkeley, Los Angeles, CA and London: University of California Press, 2001), for the efforts of the West coast Chinese community in the US to integrate in terms of health practices.
18. See Marks and Worboys, 'Introduction', 2–3.
19. See, for example, Kraut, *Silent Travelers*; Naomi Rogers, 'Dirt, Flies and Immigrants: Explaining the Epidemiology of Poliomyelitis, 1900–1916', *Journal of the History of Medicine and Allied Sciences*, 44 (1989), 486–505.
20. In terms of disease and public health, Nayan Shah has argued that exploring the category of race through the prism of public health helps illuminate how Chinese men and women (among others) were absent in the definition of both 'public' and 'health'. The health of the Chinese inhabitants of San Francisco mattered only 'instrumentally' and medical professionals represented them as a 'pestilence' (particularly in connection to smallpox, syphilis and plague), a danger to the white public, diseased and dirty. Kraut, meanwhile, has explored the blame attached to southern Europeans for epidemics and other diseases: Shah, *Contagious Divides*; Alan M. Kraut, 'Southern Italian Immigration to the United States at the Turn of the Century and the Perennial Problem of Medicalised Prejudice', in Marks and Worboys (eds), *Migrants, Minorities and Health*, 228–49.
21. See Howard Markel and Alexandra Minna Stern, 'The Foreignness of Germs: The Persistent Association of Immigrants and Disease in American Society', *Milbank Quarterly*, 80 (2002), 757–88. This is notably represented in the case of 'Typhoid Mary'. See Judith Walzer Leavitt, *Typhoid Mary: Captive to the Public's Health* (Boston, MA: Beacon Press, 1996).
22. Shah, *Contagious Divides*, 251–2.
23. Ian Robert Dowbiggin, *Keeping America Sane* (Ithica, NY and London: Cornell University Press, 1997).
24. See also John S. Haller, *Outcasts from Evolution: Scientific Attitudes to Racial Inferiority, 1859–1900* (Urbana, IL: University of Illinois Press, 1971); Daniel J. Kevles, *In the Name of Eugenics: Genetics and Uses of Human Heredity* (Berkeley, CA: University of California Press, 1985); Dowbiggin, *Keeping America Sane*.
25. Shah, *Contagious Divides*, 258.
26. Warwick Anderson, *The Cultivation of Whiteness: Science, Health, and Racial Destiny in Australia* (New York: Basic Books, 2003), 254.
27. Fairchild, *Science at the Borders*, 10.
28. Shah, *Contagious Divides*; Anderson, *The Cultivation of Whiteness*.
29. Liverpool Record Office, M614 RAI/40/2/1, Annual Report, Rainhill Asylum 1870, Report of the Medical Superintendent, 115.

1

Insanity and Immigration Restriction

Alison Bashford

In the wave of transnational scholarship on the modern regulation of global human movement, the famous immigration restriction acts in Anglophone settler colonies hold centre stage. 'Drawing the global colour line', as Marilyn Lake and Henry Reynolds have recently put it, was a core element of the great modern aspiration to produce nations out of human difference.[1] The colour line began with various Chinese exclusion acts first in California and the Australian colony of Victoria, followed by acts to regulate Indian indentured labour, to restrict Japanese entry, and to exclude, more generically, so-called 'coloured aliens' from any number of jurisdictions. This included all the Australian colonies, British Columbia, New Zealand, Natal, Newfoundland, Cape Colony and later the Union of South Africa. In the United States, the process of Asian exclusions joined a different but compounding system in the early twentieth century that limited southern and eastern European entry through a national quota system.[2]

Historians of public health have traced the connections between quarantine and immigration restriction, explaining the infectious disease rationales for exclusions and deportations.[3] It has been suggested that quarantine measures long predated modern immigration law, the legislative and bureaucratic prelude to broader regulation of movement. But in the modern period, and especially by the early twentieth century, disease prevention and the racial constitution of nations had come to be perceived as mutually constitutive in some contexts. Australia, as I argued in *Imperial Hygiene*, was aspirationally 'white' through both health and racial policy, that is, through linked quarantine and immigration restriction laws.[4] Analysis of the politics of race and ethnicity has formed the core of most scholarship at the intersection of immigration and quarantine history.

14

This chapter looks again at the overlap between immigration restriction and health, not through infectious disease management but through ubiquitous mental health and disability clauses in the immigration statutes that proliferated in the late nineteenth and early twentieth centuries. That such statutes almost always included some kind of mental health criteria of exclusion is under-recognised, both in the historiography of psychiatry and mental health and in the historiography of immigration regulation more broadly. The significant exception is Ian Dowbiggin's work on Canada and the United States, and indeed the combined history of immigration restriction and mental disability has been more strongly mapped in Canada than anywhere else.[5] And yet, by the early twentieth century almost all alien and immigration laws included a clause restricting or discouraging the entry of 'idiots or the insane', the most common descriptors used. What was the pattern of this phenomenon between the various Anglophone jurisdictions, and over time? How do we think about the insanity clauses, as separate to or as part of the powers to deport or exclude on the basis of race and ethnicity? And how, precisely, was all this part of the history of eugenics? Insanity and immigration restriction, it turns out, was foundational to the modern 'globalization of borders'.[6] This was a phenomenon that materialised earlier, and was more enduring, than exclusions on the basis of race and ethnicity; it was a transnational process shaped by racial exclusions but cannot be reduced to that.

'Idiots and the insane'

For many years, those working on the legal history of immigration, and even more so those outside the field, expected immigration acts and aliens acts to have been driven by exclusions. Amy Fairchild's work on the selective inclusion of Europeans into the US labour force through the screening process at Ellis Island, New York, began to complicate this picture. This has been followed up by Paul Kramer, who shows that Chinese 'exclusions' in the United States are rather better understood as a process by which some Chinese (merchants) were screened in, while others (labourers) were screened out.[7] The same expectation that immigration and aliens acts facilitated entry as much as dictated those who were to be turned back played out in the Antipodean colonies, notably in New Zealand. There, the first aliens act was not about exclusion, but inclusion. And the law that *did* stipulate criteria of exclusion was not directed against Chinese workers at all, but against the mentally ill. The Aliens Act (1866, amended in 1870), the separate Immigration Act

(1868) and the Imbecile Passengers Act (1873) were passed in the wake of the Maori wars.[8] Pakeha – the foreign, mainly British population – at that point numbered about 250,000. Under huge new government assistance programmes, a robust agent-general in London and immigration agents located through England, Scotland and Ireland, British and Irish emigration to New Zealand was actively promoted. Three-fifths of those who emigrated were English, one-fifth Scottish and one-fifth Irish. Persuasion to emigrate was not always an easy task, with prospective English, Scottish and Irish migrants wary of stories of the Maori wars. But the benefits of overcoming that fear were significant for individuals and families; once in New Zealand migrants could purchase land confiscated from Maori in those very wars. Travel costs were waived, and agricultural labourers and single female domestic servants were sought, provided they were sober, industrious, of good moral character and in good health. They also needed to be 'in sound mind'.[9]

The intention of the Aliens Act was to facilitate entry and to make more Pakeha out of the offspring of 'a mother being a natural-born subject of the United Kingdom' and even out of 'friendly aliens' who sought naturalisation.[10] It determined that 'alien friends' were to be treated with respect to property and inheritance rights 'as if he were a natural-born subject of Her Majesty'.[11] And the Immigration Act sought to encourage immigration 'from the United Kingdom of Great Britain and Ireland or elsewhere with the exception of the Australian Colonies'.[12] These early New Zealand immigration laws were all about bringing people in, not keeping people out, with the sole explicit exclusion initially being people from the Australian colonies, since New Zealand did not want to build its Pakeha population from convicts or ex-convicts. But with the great success of the process, exclusionary statutes soon followed. Contrary again to the expectations set up by the scholarly focus on race or ethnicity, this did not take the form of a Chinese exclusion act. Rather, the Imbecile Passengers Act was passed in 1873, the first New Zealand law specifically to nominate and define prohibited immigrants. It ordered that any owner of a ship landing with persons 'lunatic, idiotic, deaf, dumb, blind or infirm and likely to become a public charge' was to provide a bond of 100 pounds per such passenger within seven days of arrival or be charged a further penalty fine.[13] Neither convicts nor lunatics were to populate the new colony.

The specific nomination of insanity emerged, then, as quite separate to the Chinese restriction regulations that are so often taken as foundational to the history of immigration laws. This was a pattern in the early wave of immigration laws that governed movement within the British world. The Canadian Immigration Act of 1869, for example, was designed to prohibit criminals and the destitute entering from Europe, enacted just after Canadian Confederation, and deriving from quarantine regulations. It forced all vessels transporting sick or deceased passengers to report at Grosse Île, Québec. There were few other restrictions on those who could come to Canada initially, but anyone who was blind, deaf, insane or infirm was now to be recorded by the ship's captain on passenger lists.[14] In the United States, an 1891 amending act was passed that regulated the entry of all passengers other than Chinese people (whose movement was governed by different statutes), prohibiting '[a]ll idiots, insane persons; paupers or persons likely to become a public charge'.[15] This particular list of conditions was to prove resilient and, in one version or another, was to become standard.

In many jurisdictions the differing functions of immigration regulation were increasingly gathered together under one law. That is, separate labour, health and racial exclusionary acts tended to become one statute with successive clauses detailing just who was a prohibited immigrant and how this prohibition was to be implemented. This kind of catch-all immigration act was especially common in the British imperial context because of Whitehall's marked distaste for the explicit nomination of ethnicity, nationality or race. Indians, Japanese, Chinese or 'coloured aliens' were, by the Colonial Office's strong preference, not to be explicitly prohibited in law. The whole purpose of the 1897 Colonial Conference convened under Joseph Chamberlain was the diplomatic writing out of 'race' from colonial immigration law, while retaining the exclusion of coloured aliens intact in practice. The solution was contained in the so-called Natal Formula. This stipulated use of dictation tests of various kinds to exclude people without actually nominating their ethnicity: entrants were asked by customs, immigration or quarantine officials to write out a passage dictated to them, sometimes read in English, sometimes in other European languages, as a device to deliberately exclude.[16] The Natal Immigration Act of 1897 became the model for a great cluster of colonial immigration acts at the turn of the century. It also included as a prohibited immigrant any person likely to become a public charge or any idiot or insane person.[17] A suite of British Empire and Dominion Acts followed, each reading similarly,[18] and covering the colonies of Western Australia (1897) and Tasmania (1898), New Zealand (1899), the Commonwealth of Australia (1901), Canada (1902 and 1910), Hong Kong (1904), Newfoundland (1906), Fiji (1909) and the Union of South Africa (1913).[19] Each included insanity clauses as part of the new trend for immigration acts to manage entry under one statute; in effect, though not in name, the regulation of ethnicity.

There were some exceptions to this conflation of Chinese exclusion acts with broader immigration restriction measures, however. Canada, for example, broke dominion ranks and passed a separate Chinese Immigration Act in 1903.[20] It stipulated a range of exceptions for Chinese entry – students, visitors, merchants, accompanying servants might all be allowed to enter. It also stipulated the criteria of exclusion that would trump these exceptions: any person of Chinese origin who was a pauper, 'idiot or insane', had a 'loathsome disease', or who was a prostitute or who lived from the prostitution of others.[21] The Newfoundland Act Respecting the Immigration of Chinese Persons (1906) was similar, stipulating beyond ethnicity itself the exclusion of any person of Chinese origin who was 'an idiot or insane'.[22] By the same token, other statutes at the turn of the century focused on the insane specifically, without regard to race. For example, the Hong Kong Imbecile Persons Introduction Ordinance (1904) took its cue directly from the earliest New Zealand law; it was neither about race, nor about infectious disease, but specifically about insanity.[23] Likewise, the UK Aliens Act (1905) excluded a person 'if he is a lunatic or an idiot'.[24]

After the First World War, there was another cluster of laws and amendments. These were driven by the major changes in US policy; the shift from incorporating millions of European migrants from the 1890s to the controls put in place with the 1917 Act and the well-known 1924 Immigration Act. By that time, the whole question of race-based exclusions was questioned not just by Whitehall and Westminster but also far more genuinely as a matter of international law, at the Paris Peace Conference, 1919. There, a racial equality clause was put on the table by the Japanese delegation and was ultimately defeated.[25] The question of racial equality was argued largely over the Australian Immigration Restriction Act, even though a dozen or more jurisdictions had similar policies and laws (most, in fact, more explicitly racially exclusive than the Australian statute).[26] In part, because of the international delicacy of this challenge, many jurisdictions intensified their regulations, not least the prohibitions on mentally ill entrants. The United States ramped up its criteria of exclusion in the years after the Paris Peace Conference, as did Canada and Australia. New laws were passed beyond the settler colonial world as well – the Straits Settlements, for example (1919, 1932) – and by the eve of the Second World War, insanity clauses had become an entirely standard and normalised element of immigration statutes.

Asylums and public charge

What is there to say about these insanity clauses, so similar to one another, and the exclusions they made lawful? How do we explain them, or are they unremarkable, simply what we would expect to appear as part of late nineteenth- and early twentieth-century territorial national-ism that was taking a distinct biopolitical turn? The first point to note is that they were clearly and consistently about the public charge ques-tion; the expenditure of public monies. In this way, they need to be understood in the context of the faltering emergence of new kinds of welfare states in immigrant nations and colonies. In most, if not all, of these jurisdictions, insane asylums were public institutions and the cost of supporting chronically dependent and 'unproductive' entrants would be borne by unwilling receiving governments. As Dowbiggin shows with respect to Canada and the United States at the turn of the century, it was often asylum psychiatrists who spearheaded campaigns to ren-der the exclusion of the insane more stringent, typically arguing that asylum numbers needed to be kept down to manageable levels and reserved primarily for native-born populations. And for immigration restrictionists of nativist or racist bent, the disproportionate number of immigrants in asylums was constantly brought into the debate as evidence.[27] In Canadian and US contexts, anti-immigrationists pointed out census data that indicated a high proportion of the foreign-born in insane asylums across North America.[28]

Second, countries of immigration were actively shaping their national populations – at a policy level – with regard to Europe. They sought to keep a perceived Old World degeneracy out of newer countries that aspired to an improved public health. Spurious demographic and epidemiological assessment of which migrating population was more likely to end up a public charge was one constant point of intersec-tion between insanity and ethnicity. Canada's C.K. Clarke, for example, pronounced the defective and degenerate tendencies of those from Cen-tral and Southern Europe as opposed to 'the sturdy agriculturalists of the British Isles'.[29] The former were more likely to become a charge on the state. But presumptions about ethnicity were never predictable. Another Canadian medical inspector argued that the industrialised British Isles produced precisely the institution-ready population that was undesirable for Canada. Far preferable were southern Europeans, the kind of fresh labour-ready workers that Fairchild suggests the Ellis Island screening process was ultimately geared towards: '[W]e have in such races not only an industrial asset of great value but also the assurance of a population remarkably free from the degenerative effects seen in those classes which have been for several generations factory operatives and dwellers in the congested centres of large industrial populations'.[30] At times, as Alan Sears has shown, admitting the mentally ill was linked to the admittance of 'pauperism'; the undesirable creation of a dependent class. As the Canadian Board of Health put it: 'The British Poor Law has for four centuries become so integral a part of the social fabric there that immigrants brought up under its influence have, when in need or distress or sick, without hesitation drifted to the refuges, houses of industry or hospitals in Canada as naturally as they did in England'.[31] In Canada, there was a particular concern about the Barnado children emigrating, with suggestions that they were disproportionately filling asylums as well as penitentiaries, as they reached adulthood.[32]

The strong new systems and institutions of public health and welfare that were just developing in the first decade of the twentieth century – in Canada, in New Zealand, in Australia, less so in the United States – were seen to be at risk if used by newcomers, not by people who laboured locally, or whose families contributed to national or state economies. This was especially an issue for what were some of the earliest and experimental Labour governments in the world. It is for this reason that many of the acts included powers of deportation, even if an immigrant was committed to an asylum years after their original arrival; in some jurisdictions up to three years, or even up to five years later.[33] The more generous of the acts stipulated that idiots or the insane might enter if a resident, subject or citizen of the receiving country was willing to accommodate and provide all costs for support. Sometimes a large, for most a prohibitive, bond was sought – £100 for example. In all cases, the public charge issue underwrote the nineteenth-century laws and indeed continues to do so in the twenty-first century, one way or another.

The third point to note concerns the changing vocabularies of insanity. Typically the earliest laws distinguished between 'lunatic' and 'idiot', or even more commonly between 'idiot and insane': differentiating between those born without reason, and those who had lost it.[34] Occasionally, as with the early New Zealand case, the statutes spelled out deafness, dumbness and/or blindness as separate conditions again. In the early twentieth century, Canada and Australia added the 'epileptic' as a prohibited immigrant as part of the same process of refinement of categorisation, and US law excluded epileptics from 1903.[35] Importantly, the term 'feeble-minded' entered immigration law in Canada in 1906, the United States in 1907 and Australia in 1912. The *Report of the [British] Royal Commission on the Feeble Minded* in 1908 defined the 'feeble-minded' as 'persons who may be capable of earning a living under favourable circumstances, but are incapable from mental defect existing from birth or from an early age: (*a*) of competing on equal terms with their normal fellows; or (*b*) of managing themselves and their affairs with ordinary prudence'.[36] This of course raises the question of how a feeble-minded person would be recognised or diagnosed for the purposes of exclusion.[37] Compounding the problem, inspectors were charged not only with spotting the insane, but sometimes the *potentially* insane; these were the feeble-minded, any of whom, it was thought, might degenerate into clearly insane people. In the US case, for example, Ellis Island inspectors were to identify those with 'constitutional psychopathic predisposition', a hidden condition that waited only for a local trigger for the person to become actively insane and subsequently a public charge.[38]

There was a marked trend, then, for the categories of mental illness and disability to become more refined in the statutes themselves, so that any person who seemed dubious to an agent-general assessing applications for passage at point of departure, or a medical, customs or quarantine officer at point of arrival, could be prohibited entry. In Australian law, for example, the generic category of 'Insane' was dropped in 1912 and replaced by 'any idiot, imbecile, feeble minded person, or epileptic'. But this was still not specific enough. A further clause stipulated as prohibited 'any person suffering from any other disease or mental or physical defect, which from its nature is, in the opinion of an officer, liable to render the person concerned a charge upon the public or upon any public or charitable institution'. And in case that did not cover all scenarios, yet another clause was inserted: 'any person suffering from any other disease, disability, or disqualification which is prescribed'.[39] Like everything in immigration law, it was the very flexibility of the categories, especially 'feeble-minded', that was useful for the purposes of exclusion.

Eugenics, insanity and immigration restriction

The public charge problem remained steadily part of the rationale for exclusion. This did not go away. It was compounded, however, by an increasingly biological rationale for the exclusion of certain people, deriving from apparent heritability of these conditions. This made the insanity clauses more like the long-standing 'loathsome or contagious disease' clauses; rather more genuinely part of a health policy than a question of public expenditure. The new refinement of mental health disabilities in immigration law was a major manifestation of eugenics on an international scale.

Experts and authorities in countries of immigration, like Australia or the United States, would often present their nations as being in unique positions to practice eugenics via the legal infrastructure of immigration restriction: Robert DeCourcy Ward of Harvard University, to take one example.[40] Ward was a founding member of the US Immigration Restriction League. He pressed very hard for immigration to be considered both eugenically and in terms of public charge, and was enormously pleased that in the United States, unlike in Europe, there was considerable potential for the selection of citizens. He made the curious analogy to the Pilgrim Fathers: just like the first founders, immigrants of the twentieth century would be 'picked men and women'. Selection of the fathers and mothers of future American children was equally important. 'National eugenics for us means the prevention of the breeding of the unfit native, as well as the prevention of the immigration, and of the breeding after admission, of the unfit alien'.[41] That insanity was inherited was simply a fact to be incorporated into law and policy, but to make matters worse, he wrote, 'imbeciles' had larger families and larger numbers of illegitimate offspring. Taken together, this was all a 'crime against the future'.[42] For those such as Ward, insanity clauses in immigration laws were an extension of quarantine measures for the exclusion of diseased animals, of pests or of 'disease germs'. Their implementation should be intensified and made more specifically and overtly eugenic. He suggested amendments to enable the exclusion of more aliens 'of such low vitality and poor physique that they are eugenically undesirable for parenthood'.[43] Ward thought that this would constitute the real conservation of the American race. In such ways, the rationale for exclusion came to be about reproduction and inheritance in addition to the older argument of public cost.

Spencer L. Dawes, Medical Examiner of the New York State Hospital Commission and Chairman of the Inter-State Conference on Immigration, said to the American Psychiatric Association meeting in 1924 that their collective responsibility was 'to see to it that the blood stream of our country is preserved from pollution from the admixture with that of diseased and defective aliens and that the burden of taxation is made as small as is reasonably possible'.[44] Such statements from the period are entirely familiar and in many ways unsurprising, if odious. They also exemplify how by the 1920s a dual argument for the insanity clauses was typically put forward. After a long list of calculations, Dawes announced that 'the taxpayers of the State of New York are supporting more than

10,000 non-citizens, public charges in hospitals for the insane, most of whom would never have been admitted to the United States, and a considerable proportion of whom should have been removed therefrom by the Federal Government long since, had the law been enforced and its provisions observed'.[45] He noted that at Ellis Island, examinations took on average 7 seconds per person, and thought that the new quota laws had not succeeded in addressing the insanity issue at all well. One effect was 'greater laxity than ever in the examination of arriving immigrants at the ports of entry'.[46] For Dawes and many of his colleagues, the laws themselves were becoming immaterial because their implementation was lax.[47] He was one of many who argued for mental and physical examination of aliens at point of departure abroad, not on entry, and in which case the steamship company would be held accountable for the alien's qualifications for admission.

It is important to consider this triad of eugenics, insanity and immigration restriction closely. The problem of intelligence, heritability of mental conditions, intellectual disability and the so-called feeble-minded was core business for eugenicists everywhere.[48] And yet the 'eugenics' of immigration restriction is almost always interpreted as, or conflated with, racial and ethnic exclusions.[49] The historiography stresses that the national/ethnic quota system was the key outcome of this alliance of immigration restriction and eugenics.[50] This overlooks, to some extent, the eugenics of mental health exclusions on its own terms, and follows from a sometimes too-ready, or perhaps too-easy conflation of eugenics with 'race' objectives: there is a tendency to consider eugenics simply as race science in the first instance, or even to consider eugenics as *only* race science.[51] Even the very best historians do this. Alan Sears, for example, in his 1990 article on Canadian immigration restriction failed to see the mental health clauses as themselves eugenic. Rather, for Sears, it was their manifestation as race theories that made them so. He summed up that '[i]n the early twentieth century this legacy had hardened into pseudo-scientific race theories, such as those of the early twentieth century eugenics movement'.[52] It is more accurate to understand eugenics as a set of ideas about mental and physical fitness and (dis)ability in the first instance that manifested and was implemented in terms of racial difference in certain ways, in certain places. The immigration acts bear this out. What linked eugenics and immigration restriction most squarely were the insanity clauses.[53]

In the Australian context, for example, historians will typically claim that the immigration restriction act was 'eugenic' *because* it excluded coloured aliens. And yet immigration restriction was far more strictly 'eugenic' because it excluded 'unfit' (insane, idiotic, feeble-minded, deaf, epileptic) whites, almost all of whom were from the United Kingdom and Ireland. As reported in the 1920s, the most common grounds for refusal of entry was 'want of physical fitness, deficient height and weight, defective eyesight, deafness, mental deficiency, and tuberculosis'.[54] It went without saying in the 1920s that these deficient and defective would-be immigrants were British or Irish, precisely because so-called coloured aliens were already excluded. Almost entirely absent in Australian historiography is the fact that operationally it was British and Irish entrants who were most often actively excluded under the provisions of this famous immigration Act, including its insanity and mental defect clauses: whites only, but only (mentally) fit whites were admitted.[55] The rationale for *inclusion* was clearly racialised, but it was exclusion of the mentally disabled that made these laws specifically eugenic.

Insanity and ethnicity in the United States

The US situation was slightly different to the Australian, even though the mental health clauses often read similarly and sometimes even identically, across these jurisdictions. The United States never had such a dominant single stream of migrants, but always, even after the quota system, had a far more diverse immigrant population than Australia. In the United States, then, the insanity clauses and the ethnic/national quota system worked rather more explicitly in tandem. Yet it was equally about nation-building.

'May God give us strength to acquire and perpetuate the thrill of patriotism', one respondent gushed in response to Dawes' paper on 'Immigration and the Problem of the Alien Insane' at the meeting of American psychiatrists.[56] Unsurprisingly, the whole significance of immigration restriction was its relation to nationalism. But in what ways, precisely, was it about race and ethnicity? We should not presume that the response was always driven by strident race patriotism. Another respondent at this meeting cautioned, for example 'if we join an unscientific popular clamor on behalf of the so-called Nordic races we shall be ridiculed'.[57] Dawes, in fact, agreed: this was not a question of, or about, the Nordic races, he claimed, not a question of one nationality or another: 'I think we should forget all that, because this is a country composed of all kinds of races'.[58] Even studies of asylum populations did not always contain declarations against the foreign-born. H.M. Swift, assistant physician at Danvers State Hospital in Portland,

Maine noted carefully: 'it cannot be assumed that the relative frequency of insanity among the races in America is necessarily a true indicator of relative race susceptibility in residents of the mother countries, because changes in environment may have had their modifying effects'.[59] At the same time, such statements were disingenuous at the very least, since the primary correlation sought by multiple studies of asylum populations was that between ethnicity and insanity. The whole point of the major wave of epidemiological studies in the United States and Canada was to try and draw conclusions about the ethnicity of asylum populations compared to ethnicity of the insane in total populations. In most US versions, this was about Irish immigrants in the first instance and the disproportionate number of the Irish born in institutions was regularly noted.

Swift's study, like many, was quite careful. Of the foreign-born population in his state, 7.8 per cent were Irish. Of asylum populations, however, they comprised 15.8 per cent of first admissions. Once corrected for age, he concluded that if Irish adults constituted 10.2 per cent of the total population they constituted 15.8 per cent of populations of insane asylums. Swift noted that in Ireland itself the ratio of insanity in the general population was comparatively high, indeed that '[i]n the Irish we find a higher ratio of insanity than in any other people'. In 1901, 1 in every 212 people was insane, as against 1 every 309 people in England, he claimed.[60]

Swift then proceeded to study diagnoses comparatively, by ethnicity, comparing his results with George Kirby's 1909 *A Study in Race Psychopathology*.[61] He divided his asylum populations into what he called the alcoholic psychoses (acute and chronic), dementia praecox, manic-depressive insanity, general paralysis and senile and organic dementia. The first result he noted was with respect to the alcoholic psychoses that struck 9 per cent of males of native parentage, and 26 per cent of Irish parentage. Of 102 cases of alcoholic insanity (both males and females), 45 were of Irish parentage. What is notable here is that when he calculated his comparative study of diagnoses within asylums, Swift decided to shift from place of birth (the standard nomenclature 'foreign-born') to parentage. In general, he wrote, parentage 'was a more correct indication of race than nativity'.[62] The excessive use of alcohol might be 'common enough in the native stock', he concluded, but 'resistance against the establishment of a psychosis is greater'.[63] In other categories he found less difference between native American-born and Irish, and for general paralysis the frequency was less in the Irish parentage group than the native born.

Kirby, writing about the east coast, found similar results. He commented that for the Irish, the close relationship was between alcoholism, senile dementia and various organic brain diseases, whereas for the native American-born the connection was rather more between alcoholism and 'meta-syphilitic disorders such as general paralysis'. It was the American-born of Irish parentage who mainly ended up with general paralysis, according to Swift, since the foreign-born Irish (that is to say the Ireland-born Irish) were 'in general, a moral people and not prone to contract syphilis'.[64] What did this all add up to? This was Swift's last word: '[I]nsanity occurs with relatively greater frequency among the population of foreign birth and parentage than among native stock, and from this last it may be inferred that, associated with the three great causes of insanity, heredity, alcohol and syphilis, there is operative in America another potent factor in the overfilling of our public asylums, namely, immigration'.[65] The borders needed closer scrutiny for the insane, who, such studies suggested, were more likely to be found in one ethnic group over another.

Thomas Salmon, in 1907, thought that 'the prevalence of insanity among the Irish in the United States has no parallel in the world'. One per 203 persons institutionalised in Ireland became one in every 121 persons in the United States. But something similar was going on with the English too: one in 209 persons institutionalised in New York State, compared to one in 288 in England.[66] This is partly why Salmon wanted an alternative process to immigrant inspection. He suggested a catch-all test for illiteracy that would exclude many of the insane instantly (as well as many others), and as he would have it, far more easily.[67] That is, this would 'diagnose' the insane, the potentially insane and the feeble-minded whatever their nationality or ethnicity more readily and effectively than individual inspections. While authorities took note of this suggestion, the practice at entry points like Ellis Island remained one of individual medical inspection.

This led, in principle if not in practice, to the idea that diagnostic questions should also be culturally specific. Culture and ethnicity – as well as the alien's fatigue, excitement or nervousness – should be taken into consideration. New diagnostic performance tests were developed for Ellis Island officers' use that in part 'allowed for' ethnicity: 'The Imbecile Test', 'The Moron Test', as well as an Ink-Blot imagination test.[68] At the end of the day, wrote Ellis Island inspector Howard Knox, the detection of the 'moron or higher defective' was vastly more important than the detection of the 'insane': the latter, he thought, would come to be recognised soon enough and would find their way into an asylum anyway and might thereafter be deported.[69]

C.P. Knight, assistant surgeon at Ellis Island, constantly stressed the difficulty in distinguishing between the insane and the feeble-minded. It was critical to do so, he thought, since census statistics showed that 30 per cent of the feeble-minded children in the general population of the United States 'are the progeny of aliens or naturalized citizens'. This class is 'highly prolific', he said. Knight advised that the idiot is easily and quickly recognised visually: low receding forehead; disproportionately large face with respect to cranium; nose too large or too small, or deviated, or flat; excessively deep orbits; bad teeth; arching palate; the skin 'mongolian coloration or albino'. Cretenism, he said, was equally easy to determine. The imbecile, though, needed to be recognised more through speech and the feeble-minded was the most difficult of all to isolate, because the problems were largely cognitive.[70]

Interestingly, Knight insisted that all of this mental capacity and incapacity could only be pinpointed vis-à-vis ethnicity. 'An officer with experience becoming familiar with the different races, studying closely their characteristics, knowing something of their language, can tell at a glance the abnormal from the normal as they pass him on the line'. The examiner needed to know 'the mean type of the race' and its deviations, by gait, stature, and expression. 'The close application to the study of the race is more important in the determination of the mental status of the alien than in the diagnosis of physical abnormalities'. This involved determining normal conduct for that race, and assessing the individual in that light. 'It is perfectly normal for the southern Italian to show emotion on the slightest provocation but should he show the stolidity and indifference of the Pole or Russian, we would look on him with suspicion and perhaps hold him for a detailed examination'. The English and German immigrants answer questions promptly, 'but should they become evasive as do the Hebrews, we would be inclined to question their sanity'.[71] The system expected and allowed for racial/cultural difference. A good examiner needed to be able to comprehend a thoroughly normal Italian, Greek or Pole before even hoping to recognise a mentally defective one. In other words, the alienist in the United States needed to know ethnicity before he could know insanity.

Continuity: The long twentieth century

Clearly it is important to complicate our understanding of exactly how ethnicity and insanity functioned with respect to one another, in the context of immigration regulation. A final reason to do so concerns periodisation. The historiographical focus on race and ethnicity has meant that the end of race-based exclusions has implied the end of immigration restriction. The repeal of devices like the literacy or dictation tests (in Australia in 1958 for example), the end of explicit racial nominations as in the Canadian Immigration Act, or the end of the US quota system in 1965, for example, often round out historians' analysis of immigration restriction. And yet this does not accord with the history of the acts themselves. Typically, while race and ethnicity were gradually dropped as criteria for exclusion during the 1950s and 60s, much else remained. This was, and is, certainly the case with respect to mental ill-health. And entirely new migration acts and mental health clauses were often created. In Hong Kong, for example, a 1949 ordinance was enacted 'to control the population of the Colony by providing for the expulsion of undesirables therefrom'. 'Undesirables' included any person who was without means of subsistence and was diseased, maimed, blind, idiot, lunatic or decrepit; persons likely to become a vagrant, beggar or a public charge. Undesirables would be accommodated in camps prior to their expulsion from the Colony, as would be 'suspected undesirables'.[72] In Australia, the infamous Immigration Restriction Act was importantly replaced by the Migration Act in 1958, the beginning of the end of the white Australia policy, but mental health clauses remained intact. If one became an inmate of a mental hospital within five years after arrival, deportation was lawful. The 'prescribed diseases' that made a person a prohibited immigrant included a physical or mental disability or defect.[73] And in the major 1992 Australian overhaul of migration law, the 'health criterion' required for a visa retained specified physical or mental conditions.[74] Currently, the health requirement for intending immigrants is expressed as having a threefold purpose: to minimise public health and safety risks to the Australian community; to contain public expenditure on health and community services, including Australian social security benefits, allowances and pensions; and to maintain access of Australian residents to health and other community services. It is stipulated that '[I]n line with Australia's global non-discriminatory immigration policy, the health requirement applies equally to all applicants from all countries, although the extent of testing will vary according to the circumstances of each applicant'.[75]

In Canada, a 1952 law took a different tack and became not more general but more specific. The prohibited class included persons who were idiots, imbeciles or morons, were insane or had been insane at any time, had constitutional psychopathic personalities or were afflicted with epilepsy. Immigrants who were dumb, blind or otherwise physically defective were prohibited from landing unless they were unlikely to become public charges or they already had family in Canada. Criminals, prostitutes, homosexuals, pimps, procurers, professional beggars and vagrants, chronic alcoholics, drug addicts, drug pedlars, members of subversive organisations, spies, saboteurs, persons found guilty of espionage or treason were all prohibited from Canada. This was in addition to the diseased, of course. Any person entering Canada might be mandatorily examined (mentally or physically or both) by a medical officer. Additionally, anyone 'mentally or physically abnormal to such a degree as to impair seriously their ability to earn a living' were prohibited immigrants. Exceptions were made for the entry of people under (private) treatment and care at a health resort, hospital, sanitarium or asylum.[76] A 1976 Canadian Act removed many of the restrictions placed on the immigration of people with mental or physical disabilities and provided the framework for current immigration policy. Potential immigrants to Canada are now separated into three classes: family class, composed of immediate family of Canadian citizens or residents; humanitarian class, which introduced refugees who fit the official UN description, as well as persecuted or displaced people; and independent class, who apply for landed immigration status on their own and must go through selection based on a points system. In fact, in all classes, Canadian law has returned to a broad catch-all prohibition of those likely to become a burden on social welfare or services.[77]

In the United States, the Chinese Exclusion Acts were repealed in 1943. And the extensive 1952 Immigration and Nationality Act abolished the 1917 Asian Barred Zone and allowed immigration into the United States based on strict ethnic and numerical quotas. This winding down process is often seen to have been completed by important 1965 amendments, which in essence removed 'natural origins' as the basis of American immigration legislation, stating: 'No person shall receive any preference or priority or be discriminated against in the issuance of an immigrant visa because of his race, sex, nationality, place of birth, or place of residence'. So far, so good, but there was a literal qualification: 'Except as specifically provided'. Mental health exceptions were retained and in some instances were extended. The words 'mentally retarded' replaced 'feeble minded'. Epilepsy was removed as a category, but substituted with the words 'or sexual deviation'. There were, then, specific provisions regarding the following: 'persons (1) mentally retarded, (2) insane, (3) afflicted with psychopathic personality, or with sexual deviation, (4) a chronic alcoholic, (5) afflicted with any dangerous contagious disease, or (6) a narcotic drug addict'.[78]

In terms of periodising immigration restriction, the United Kingdom stands alone. It belatedly, and very reluctantly, joined comparable jurisdictions with the Aliens Act (1905). Conversely, in the 1960s, just when the other nations were undoing their legislative ties to nationality, race and ethnicity, the United Kingdom was trying to figure out ways to tighten such controls through the 1962 and 1968 Commonwealth Immigrants Acts. The 1962 Act prohibited an immigrant 'if it appears to the immigration officer on the advice of a medical inspector ... that he is a person suffering from mental disorder, or that it is otherwise undesirable for medical reasons that he should be admitted'.[79] In the new 1968 Act, which aimed to facilitate immigration from the 'white' dominions of the Commonwealth while retaining mechanisms to exclude people from other parts of the Commonwealth, this mental health provision was removed.

Conclusion

Experts and authorities in countries of immigration, like Australia or the United States, would often boast of the possibilities for strongly shaping the character and health of their national populations, present and future, via immigration screening. This contrasted strongly with emigrant countries.[80] They were correct to identify the potential of the immigration restriction processes. By the early twentieth century, an entire hemisphere was implementing such laws. The exclusion of insane foreigners, especially from new world states that were experimenting with health and welfare measures, became part of the business of the state, as well as the business of the emergent discipline of psychiatry. The consensus about exclusion was almost total, across all jurisdictions, based on public cost rationales and eugenic concerns. The links with immigration made experts on insanity also would-be experts on ethnicity.

By focusing on the longevity of mental health criteria, not just the continuity but the global normalisation of immigration regulation becomes evident. Over time, immigration restriction became a universal requirement of all nations; perhaps the key expression of sovereign independence in a globalised world. In this process, the nomination of race, nationality or ethnicity as criteria for exclusion rose to prominence, was critiqued and ultimately became internationally unacceptable. Mental health exclusions, by contrast, often predated such criteria, were retained in modern migration law and remain in operation in many instances. The critique of racial discrimination never transferred successfully to a human rights–based critique of discrimination against the mentally ill, despite a phenomenally successful public critique of eugenics in other spheres. The histories of the alien and the alienist are linked.

Notes

I am grateful for the research assistance of Annie Briggs, Catie Gilchrist and Chis Holdridge. This chapter forms part of the Australian Research Council project, 'Immigration Restriction and the Racial State', DP0984518.

1. M. Lake and H. Reynolds, *Drawing the Global Colour Line: White Men's Countries and the Question of Racial Equality* (Melbourne: Melbourne University Press, 2008).
2. A. Zolberg, *A Nation by Design: Immigration Policy in the Fashioning of America* (Cambridge, MA: Harvard University Press, 2006); A. McKeown, *Melancholy Order: Asian Migration and the Globalization of Borders* (New York: Columbia University Press, 2008).
3. For example, A.M. Kraut, *Silent Travellers: Germs, Genes and the 'Immigrant Menace'* (Baltimore, MD and London: John Hopkins University Press, 1995); H. Markel, *Quarantine! East European Jewish Immigrants and the New York City Epidemics of 1892* (Baltimore, MD and London: Johns Hopkins University Press, 1997); A.L. Fairchild, *Science at the Borders: Immigrant Medical Inspection and the Shaping of the Modern Industrial Labor Force* (Baltimore, MD and London: Johns Hopkins University Press, 2004).
4. A. Bashford, *Imperial Hygiene: A Critical History of Colonialism, Nationalism and Public Health* (Houndmills: Palgrave Macmillan, 2004).
5. I.R. Dowbiggin, *Keeping America Sane: Psychiatry and Eugenics in the United States and Canada, 1880–1940* (Ithaca, NY and London: Cornell University Press, 1997). See also, T.D. Comeay and A.L. Allahar, 'Forming Canada's Ethnoracial Identity: Psychiatry and the History of Immigration Practices', *Identity*, 1:2 (2009), 143–60; A. Sears, 'Immigration Controls as Social Policy: The Case of Canadian Medical Inspection 1900–1920', *Studies in Political Economy*, 33 (1990), 91–112.
6. The phrase is Adam McKeown's in *Melancholy Order*.
7. Fairchild, *Science at the Borders*; P.A. Kramer, 'Empire against Exclusion in Early 20th Century Trans-Pacific History', *Nanzan Review of American Studies*, 33 (2011), 13–32.
8. J. Belich, *The Victorian Interpretation of Racial Conflict: The Maori, the British, and the New Zealand Wars* (Montreal: McGill-Queen's University Press, 1989), 258–88.
9. R.M. Dalziel, *The Origins of New Zealand Diplomacy: The Agent-General in London, 1870–1905* (Wellington: Victoria University Press, 1975), 36.
10. New Zealand, *Aliens Act* (1866), No. XVII, s. 2.
11. New Zealand, *Aliens Act* (1870), No. XL, s. 2.
12. New Zealand, *Immigration Act* (1868), No. XLII, s. 2.
13. New Zealand, *Imbecile Passengers Act* (1873), No. LXX, s. 3.
14. Canada, *Immigration Act* (1869).
15. United States of America, *An act in amendment to the various acts relative to immigration and the importation of aliens under contract or agreement to perform labor* (1891), Chapter 551, s. 1.

16. M. Lake, 'From Mississippi to Melbourne via Natal: The Invention of the Literacy Test as a Technology of Racial Exclusion', in A. Curthoys and M. Lake (eds), *Connected Worlds: History in Transnational Perspective* (Canberra: ANU E Press, 2005), 209–30.

17. J. Martens, 'A Transnational History of Immigration Restriction: Natal and New South Wales, 1896–97', *Journal of Imperial and Commonwealth History*, 34:3 (2006), 323–44.

18. A. Bashford and C. Gilchrist, 'The Colonial History of the 1905 Aliens Act', *Journal of Imperial and Commonwealth History*, 40:3 (2012), 409–27.

19. Canada, *Immigration Act* (1910). Chapter 27, s. 3: prohibited classes included 'idiots, imbeciles, feeble-minded persons, epileptics, insane persons, and persons who have been insane within five years previous'. Union of South Africa, *Immigration Act* (1913), No. 22, s. 4: prohibited classes included 'any idiot or epileptic or any person who is insane or mentally deficient, or any person who is deaf and dumb, or deaf and blind, or dumb and blind, or otherwise physically afflicted, unless in any such case he or a person accompanying him or some other person give security to the satisfaction of the Minister for his permanent support in the Union, or for his removal therefrom whenever required by the Minister'.

20. Canada, *The Chinese Immigration Act* (1903), Chapter 8.

21. Ibid.

22. Newfoundland, *An Act Respecting the Immigration of Chinese Persons* (1906), Chapter 2, s. 5.

23. Hong Kong, *Imbecile Persons Introduction Ordinance* (1904), No. 1.

24. United Kingdom, *Aliens Act* (1905), Chapter 13, s. 3. See also Bashford and Gilchrist, 'The Colonial History of the 1905 Aliens Act'.

25. N. Shimazu, *Japan, Race and Equality: The Racial Equality Proposal of 1919* (London: Routledge, 1999).

26. S. Brawley, *The White Peril: Foreign Relations and Asian Immigration to Australasia and North America, 1919–1978* (Sydney: University of New South Wales Press, 1995).

27. Dowbiggin, *Keeping America Sane*, 144, 195.

28. See B.A. Locke, M.S. Morton Kramer, and N. Pasamanick, 'Immigration and Insanity', *Public Health Reports*, 75:4 (1960), 301.

29. Rockwood Lunatic Asylum, 'Annual Report of the Medical Superintendent of the Asylum for the Insane, Kingston, for the Year ending 30 September, 1903', in *Thirty-Sixth Annual Report of the Inspector of Prisons and Public Charities upon the Lunatic and Idiot Asylums of the Province of Ontario, being for the Year ending 30th September 1903* (Toronto: Queen's Printer, 1904). Cited in Dowbiggin, *Keeping America Sane*, 142.

30. P.H. Bryce, 'Report of the Chief Medical Officer', in Canada, Department of the Interior, *Annual Report, Immigration*, 1909–10 (Ottawa, 1910), 110. Cited in Sears, 'Immigration Controls as Social Policy', 101. See Fairchild, *Science at the Borders, passim*.

31. P.H. Bryce, 'Report of the Chief Medical Officer', in Canada, Department of the Interior, *Annual Report, Immigration*, 1910–11 (Ottawa, 1911), 127. Cited in Sears, 'Immigration Controls as Social Policy', 100.

32. Dowbiggin, *Keeping America Sane*, 140.

33. P. Martyr, 'Having a Clean Up? Deporting Lunatic Migrants from Western Australia, 1924–1939', *History Compass*, 9:3 (2011), 171–99.

34. H.J. Stephen, *New Commentaries on the Laws of England*, 1874, vol. 2, 62. See also A. Digby and D. Wright (eds), *From Idiocy to Mental Deficiency: Historical Perspectives on People with Learning Difficulties* (London and New York: Routledge, 1994).

35. Canada, *Immigration Act* (1906). Chapter 27, s. 26. 'No immigrant shall be permitted to land in Canada, who is feeble-minded, an idiot, or an epileptic, or who is insane, or has had an attack of insanity within five years; nor shall any immigrant be so landed who is deaf and dumb, or dumb, blind or infirm unless he belongs to a family accompanying him or already in Canada and which gives security, satisfactory to the Minister, and in conformity with the regulations in that behalf, if any, for his permanent support if admitted into Canada'.

36. British Parliamentary Papers, Report of the Royal Commission on the Care and Control of the Feeble Minded, 1908 (Cd. 4202) XXXIX, 159.

37. See descriptions of this process and its difficulties in Dowbiggin, *Keeping America Sane*, 203–4.

38. Howard A. Knox, 'The Moron and the Study of Alien Detectives', *Journal of the American Medical Association*, 60:2 (1913), 105.

39. Commonwealth of Australia, *Immigration Amendment Act* (1912), No. 38, s. 3.

40. For example, Robert DeC. Ward, 'Our Immigration Laws from the Viewpoint of Eugenics', *American Breeders Magazine*, 4 (1912), 20.

41. Ibid., 21.

42. Ibid., 22.

43. Ibid., 24.

44. Spencer L. Dawes, 'Immigration and the Problem of the Alien Insane', *American Journal of Psychiatry*, 81:3 (1925), 450.

45. Ibid., 451.

46. Ibid., 457.

47. See, for example, Frank B. Hall, 'Discussion', in Dawes, 'Immigration and the Problem of the Alien Insane', 464; Thomas W. Salmon, *Insanity and the Immigration Law* (Utica, NY: State Hospitals Press, 1911), 6–7.

48. M. Thomson, *The Problem of Mental Deficiency: Eugenics, Democracy and Social Policy in Britain, 1870–1939* (Oxford: Oxford University Press, 1998); idem, 'Eugenics, Disability, and Psychiatry', in Alison Bashford and Philippa Levine (eds), *The Oxford Handbook of the History of Eugenics* (Oxford: Oxford University Press, 2010), 116–33.

49. The connection between the Immigration Restriction League and the American Breeders' Association was tight, and the Eugenic Record Office's Harry Laughlin's role as an expert witness for the House Committee on Immigration and Naturalization in 1920 was clear and significant. E. Barkan, 'Reevaluating Progressive Eugenics: Herbert Spencer Jennings and the 1924 Immigration Legislation', *Journal of the History of Biology*, 24:1 (1991), 91–112; D. King, *Making Americans: Immigration, Race, and the Origins of the Diverse Democracy* (Cambridge, MA: Harvard University Press, 2000); A. McKeown, 'Ritualization of Regulation: The Enforcement of Chinese Exclusion in the United States and China', *American Historical Review*, 108 (2003), 377–403;

J.-P. Beaud and J.-G. Prevost, 'Immigration, Eugenics and Statistics: Measuring Racial Origins in Canada (1921–1941)', *Canadian Ethnic Studies*, 28:2 (1996), 1–25; K.M. Ludmerer, 'Genetics, Eugenics, and the Immigration Restriction Act of 1924', *Bulletin of the History of Medicine*, 46:1 (1972), 59–81; R.F. Zeidel, *Immigrants, Progressives, and Exclusion Politics: The Dillingham Commission, 1900–1927* (DeKalb, IL: Northern Illinois University Press, 2004). But see also S.-T. Ly and P. Weil, 'The Antiracist Origin of the Quota System', *Social Research*, 77:1 (2010), 45–78.

50. The quota system emerged first as an emergency measure and subsequently (1924) as the Immigration Act; by which the number of immigrants by national origin was determined and limited as a percentage of the 1890 population, although Dowbiggin argues that US psychiatrists' theories of immigration need to be distinguished from the US nativists (as in the Immigration Restriction League). Dowbiggin, *Keeping America Sane*, 192.

51. See A. Bashford, 'Where Did Eugenics Go?', in Bashford and Levine (eds), *The Oxford Handbook of the History of Eugenics*, 539–58.

52. Sears, 'Immigration Controls as Social Policy', 105.

53. A.J. Rosanoff, 'Some Neglected Phases of Immigration in Relation to Insanity', *American Journal of Psychiatry* (July 1915), 45–58; Thomas W. Salmon, 'The Relation of Immigration to the Prevalence of Insanity', *American Journal of Psychiatry* (July 1907), 53–71; H.L. Reed, 'Immigration and Insanity', *The Journal of Political Economy*, 21:10 (1913), 954–6; P.H. Bryce, 'Insanity in Immigrants', *American Journal of Public Hygiene* (1910), 146–54; Knox, 'The Moron and the Study of Alien Defectives', 105–6; Ward 'Our Immigration Laws from the Viewpoint of Eugenics', 20–6. Thomas Salmon, Chair of the Board of Alienists under the State Commission in Lunacy, for example, wanted examination at the ports of departure to supplement arrival examinations. Salmon, *Insanity and the Immigration Law*, 4.

54. W.E. Agar, 'Some Eugenic Aspects of Australian Population Problems', in P.D. Phillips and G.L. Wood (eds), *The Peopling of Australia* (Melbourne: Macmillan, 1928), 142.

55. See Bashford, *Imperial Hygiene*, 152–3.

56. Hall, 'Discussion', in Dawes, 'Immigration and the Problem of the Alien Insane', 465.

57. Williams, 'Discussion', in Dawes, 'Immigration and the Problem of the Alien Insane', 467.

58. Dawes, 'Immigration and the Problem of the Alien Insane', 469.

59. H.M. Swift, 'Insanity and Race', *American Journal of Psychiatry*, 70:1 (1913), 143.

60. Swift, 'Insanity and Race', 146–7. See Cox, Marland and York's chapter in this volume.

61. George H. Kirby, *A Study in Race Psychopathology* (New York State Hospital Bulletins, 1909).

62. Swift, 'Insanity and Race', 149.

63. Ibid., 150.

64. Kirby, *A Study in Race Psychopathology*. Cited in Swift, 'Insanity and Race', 151.

65. Swift, 'Insanity and Race', 154.

66. Thomas W. Salmon, 'The Relation of Immigration to the Prevalence of Insanity', *American Journal of Psychiatry* (July 1907), 63.

67. Salmon, *Insanity and the Immigration Law*, 8.
68. J.T.E. Richardson, 'Howard Andrew Knox and the Origins of Performance Testing on Ellis Island, 1912–1916', *History of Psychology*, 6:2 (2003), 153.
69. Knox, 'The Moron and the Study of Alien Defectives', 105–6.
70. C.P. Knight, 'The Detection of the Mentally Defective Among Immigrants', *Journal of the American Medical Association*, 60:2 (1913), 106–7.
71. Ibid.
72. Hong Kong, *The Expulsion of Undesirables Ordinance* (1949), Chapter 242, s. 4, s. 6.
73. Commonwealth of Australia, *Migration Act* (1958), No. 62, s. 13, s. 16.
74. Commonwealth of Australia, *Migration Reform Act* (1992), No. 184.
75. Commonwealth of Australia, Fact Sheet 22 – The Health Requirement, immi.gov.au/media/fact-sheets/22health.htm (accessed 1 November 2011).
76. Canada, *Immigration Act* (1952), Chapter 325, s. 5.
77. Canada, *Immigration Act* (1976), Chapter 52, s. 19.
78. United States of America, *An Act to amend the Immigration and Nationality Act, and for other purposes* (1965), Public Law 89–236, s. 15, s. 18.
79. United Kingdom, *Commonwealth Immigrants Act* (1962), Chapter 21, s. 2.
80. For example, Ward, 'Our Immigration Laws from the Viewpoint of Eugenics', 20.

2
Itineraries and Experiences of Insanity: Irish Migration and the Management of Mental Illness in Nineteenth-Century Lancashire

Catherine Cox, Hilary Marland and Sarah York

The Irish and Irish migrants have been depicted as particularly prone to mental illness and institutionalisation in the nineteenth and twentieth centuries on a global scale.[1] Yet there has been only limited work on this phenomenon in the case of Irish migration to England.[2] Across the period of our study, the relationship of Ireland, part of the British state until 1922, as well as Irish migrants' relationship with the British Empire has been contested and debated.[3] Irish migrant labourers were regarded as an important and necessary resource for the British Empire and Industrial Revolution. Throughout the nineteenth century – in comparison with the situation outlined in Alison Bashford's chapter – there was a largely unmanaged movement of Irish migrants from Ireland to Britain, unchecked by immigration controls or medical examinations.[4] Yet this migration prompted anxieties in terms of welfare provision and the obligations of the English Poor Law to provide for Irish paupers, and the Irish were blamed for exacerbating already dire conditions in many communities and for spreading disease. Large numbers of Irish migrants would also end up as patients in England's growing asylum system, where, as in many other parts of the British Empire, they made up one of the largest – and in many years the largest – ethnic group.

This chapter offers an analysis of issues emerging from research into the migratory patterns of Irish patients through the Lancashire asylum system in the mid- to late nineteenth century. Arriving in a state of optimism or more often extreme distress at the port of Liverpool, one of the main ports of entry from Ireland, large numbers of Irish migrants found their way into Lancashire's four large public asylums where many remained for long periods. While Irish migration into Lancashire was firmly established at the start of the nineteenth century, the city of Liverpool – notably its Poor Law authorities and ratepayers and public health provision – was overwhelmed by the influx of Famine Irish after 1846. The impact of the Irish on the Lancashire asylum system was also enormous and enduring. What began largely as part of a story of an emergency response to Famine migration and its associated distress segued in the late nineteenth century into concern about the links between the Irish and degeneration, chronicity and criminality, all of which were emphasised in asylum records.

Here we will explore attempts to manage and cope with the stream of Irish admissions into the Lancashire asylum system in the post-Famine decades. Drawing on annual reports, admission registers and certificates, case histories and the reports of the Commissioners in Lunacy, the impact of Irish migration on Lancashire asylums will be considered against the broader backdrop of debates about the effects of Irish migration on Lancashire in general and Liverpool in particular, as evidenced in Poor Law and public health records and the local press. Asylum management regimes, shaped by the approaches of moral therapy and non-restraint, ensured that maintaining order and regularity preoccupied asylum officials throughout the nineteenth century.[5] This was challenged as asylums silted up with chronic 'incurable' patients as well as patients who confounded the ideals of moral therapy, which centred on establishing orderly behaviour and routines of regular mealtimes, work, exercise and rest. The huge number of Irish admissions contributed greatly to these problems, not only stretching asylum resources, but the characteristics associated with Irish patients – poor physical health and disruptive behaviour – further taxed the institutions' capacities to manage this patient group within the strictures of moral management in severely overcrowded conditions. In addition, the cost of maintaining these patients, particularly given their tendency to remain in the asylum for many years, represented a significant drain on local rates, fuelling anti-Irish sentiment.

Migration to Lancashire

By the middle of the nineteenth century, Irish migration into Lancashire had seen the creation of a number of established Irish communities, encouraged by the need for Irish labour in the textile and other industries. During the demographic and humanitarian disaster of the Great Famine (1846–51) and in the years that followed the number of Irish emigrants to England escalated dramatically.[6] The Famine led to over

1 million Irish deaths from starvation and disease, and around 2 million people left Ireland during the Famine and in its immediate after-math, the majority of whom migrated to Britain, the United States and Canada, and smaller numbers to New South Wales and New Zealand.[7] Estimates vary hugely, but Liverpool, ahead of other major port cities, including London and Bristol, bore the brunt of migration into England, and other towns and cities in Lancashire also saw a rapid growth in Irish inhabitants. One estimate put the number arriving in Liverpool in the first half of 1847 alone at just under 300,000, of whom around 40 per cent were paupers.[8] Frank Neal has estimated that in 1841, 17 per cent of the total population of the city was Irish-born and by the 1851 census, this figure had reached 22.3 per cent.[9]

The nineteenth century was marked by considerable variation in migratory patterns from Ireland. Fuelled by steamship companies only too willing to provide transportation, it was a relatively cheap and easy process to cross the Irish Sea. Many travelled as seasonal labour-ers and, even during the Famine and post-Famine period, though most left Ireland out of desperation, others intended to start a new life in England or North America. Large numbers became trapped by poverty, illness or other misfortunes in Liverpool.[10] Most were unskilled, Roman Catholic paupers who travelled from the rural Irish midlands and arrived in Liverpool impoverished and starving, their weak state of health fur-ther undermined by the appalling conditions on board the steamships bringing them from Ireland.[11] While in the first half of the nine-teenth century many Irish migrated in family units, after the 1860s the young and single came to dominate, individuals lacking family support networks and highly vulnerable during trade depressions. Seen as impor-tant sources of labour for local industries at times of peak employment and willing to take on work that no one else wanted, including heavy labouring jobs, during economic downturns the Irish were depicted as depressing wages and as strike-breakers. Though there was increasing diversity in the class of migrant, by the 1870s and 1880s the majority of the Irish continued to be employed as unskilled labourers working long hours for poor wages and were especially vulnerable to poverty and unemployment.[12]

Welfare, public health and disease

The arrival of Irish Famine migrants triggered shock and alarm among Liverpool's citizens, concerned about their dreadful suffering as well as the impact on the public purse. The editor of the *Liverpool Mercury* commented that 'The number of starving Irish – men, women and children – daily landed on our quays, is appalling and the parish of Liverpool has at present the painful and most costly task to encounter, of keeping them alive – if possible'.[13] The Liverpool authorities rapidly developed a series of emergency measures to cope with the influx, opening up large receiving sheds on the dockside, which attempted to provide food and shelter. Though some migrants were able to find low-paid employment or moved on, the majority who remained in the city were to become dependent on poor relief. By late 1846, the number of Irish paupers receiving public assistance in Liverpool had reached 13,471 compared with 900 in 1845.[14] Writing to the Home Secretary in 1849, Liverpool Magistrate, Edward Rushton, reported how 296,231 persons landed in Liverpool from Ireland in 1847; over 110,000 were half-naked and starving paupers, who, immediately on landing, became applicants for parochial relief.[15] By 1847–48, 49 per cent of all Liverpool Select Vestry's outdoor relief, some £20,750, was spent on Irish migrants.[16]

Though obliged to provide relief to Irish migrants, the Liverpool authorities were entitled to use removal and settlement legislation to return them to Ireland. However, changes to the legislation introduced in 1845 and 1846, coupled with the extreme conditions in Liverpool during the Famine, made it impossible to enforce the acts.[17] After 1846 Rushton seldom implemented the legislation, though Charles Willmer, a member of the Liverpool Select Vestry, insisted in 1851 that non-implementation cost the parish an additional £15,000 annually, as paupers, 'principally Irish', claimed to have residency in Liverpool.[18] While Willmer doubted the authenticity of most cases, he added that it was not possible to investigate them 'in a dense population like that of Liverpool'.[19] Neal has also suggested that some Irish migrants stopped claiming relief to avoid removal and instead remained begging in Liverpool.[20]

The Irish settling in the city by and large became concentrated in overcrowded, unsanitary and disadvantaged areas in Liverpool's worst streets, lodging houses and cellars where they became associated with the spread of disease. Even before their arrival, Liverpool's slums were notorious for their appalling conditions and the arrival of Irish migrants in large numbers made matters much worse. *The Times* reported in 1847 that

> Ireland is pouring into the cities, ... a fetid mass of famine, nakedness and dirt and fever. Liverpool, whose proximity to Ireland has already procured for it the unhappy distinction of being the most unhealthy town in this island, seems destined to become one mass of disease.[21]

Though the Famine influx was unprecedented, the experience of managing the emergency shaped Liverpool's attitudes towards Irish migrants for decades to come. The Irish had long been held responsible for outbreaks of epidemic disease in Lancashire's towns and cities, and this association persisted during the epidemics of cholera, typhus, smallpox and measles, which raged through Liverpool in the 1860s and 70s. In 1844 William Henry Duncan, subsequently appointed as Liverpool's Medical Officer of Health in 1847 (the first such post in England), reported how the Irish poor were especially exposed to fever: 'It is they who inhabit the filthiest and worst-ventilated courts and cellars, who congregate the most numerously in dirty lodging-houses, who are the least cleanly in their habits, and the most apathetic about everything that befalls them.'[22] Irish workhouse inmates were also implicated in the outbreak of smallpox in 1871–72 when the removal of Irish paupers back to Ireland was suspended.[23] Crowded into the worst areas of town, including the notorious Irish lodging houses, the Medical Officers of Health closely monitored the residential patterns of the Irish and their habits and practices. One source of cholera infection during the 1866 outbreak was identified as the wake of Mrs Boyle of Bispham Street at whose funeral the neighbours engaged in the 'incautious orgies' that 'still linger as dregs of ancient manners among the funereal customs of the Irish peasantry'.[24] Thus outbreaks of disease were associated not just with the grim conditions under which many poor Irish lived, but also with their cultural practices and behaviour.

Migration into the Lancashire asylums

It is against this background of broader health and welfare concerns that the Irish began to make a huge impact in another area of health provision, the management of mental illness. Large numbers of Irish-born patients found their way into Lancashire's four public asylums at Lancaster Moor (established 1816), Rainhill (1851), Prestwich (1851) and Whittingham (1873). Many remained for years, even decades, until their deaths from disease or old age. Their impact on the Lancashire asylums was enormous; by the late 1850s they accounted for half the admissions to Liverpool's Rainhill Asylum (see Figure 2.1) and by 1871 they made up half of the resident population.

Irish patients were not the only migrant group admitted to Rainhill and other Lancashire asylums; patient intake increasingly reflected the more diverse and globalised nature of movement in and out of Liverpool by the late nineteenth century (Table 2.1). Persistently high numbers

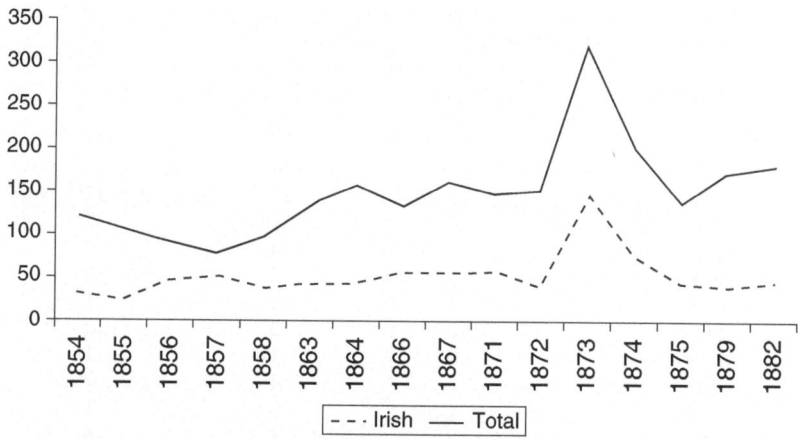

Figure 2.1 Number of Irish and non-Irish patients admitted to Rainhill Asylum, 1854–1882
Source: Annual Reports, Rainhill Asylum, 1854–1882.

Table 2.1 Table showing the countries to which patients admitted to Rainhill Asylum during 1856 belong

Country	Male	Female	Total
English	25	24	49
Irish	13	20	33
Welsh	2	4	6
Scotch	1	1	2
Manx	0	1	1
Total	41	50	91

Source: Wellcome Library, Annual Report, Rainhill Asylum 1856, Report of the Resident Medical Officer and Superintendent, Table VI, 95.

of Scottish and Welsh patients were admitted, and by the 1880s, as Table 2.2 demonstrates, Rainhill took in patients from Europe, America and Australia. Nonetheless, after English admissions, Irish-born patients made up the largest ethnic cohort for the remainder of the century, contributing to the overcrowding of what was to become the one of the largest asylum systems in England and, indeed, worldwide. Paralleling the declining impact of the Irish on public health and welfare more broadly, the number of Irish diminished in terms of overall admissions over the course of the century. Yet they remained a persistent and serious drain on the asylums' resources. They also – much more so than other

Table 2.2   Table showing the countries to which patients belong who were in Rainhill Asylum on 31 December 1888

Country	Male	Female	Total
English	372	421	793
Irish	132	124	256
Welsh	10	27	37
Scotch	14	21	35
Manx	1	4	5
German	4	5	9
Austrian	1	0	1
American	0	2	2
Italian	0	1	1
Scandanavian	3	4	7
Swiss	1	0	1
Polish	3	3	6
Others	1	2	3
Unknown	0	1	1
Total	542	615	1,157

Source: Liverpool Record Office M614 RAI/40/2/6, Annual Report, Rainhill Asylum, 1888, Table XVI, 133.

groups of patients – became entrenched in the system. Compared with non-Irish patients, they remained in the asylum for prolonged periods, exhibited poor recovery rates, contributed to the asylum's 'silting up' with chronic patients and to high mortality rates.

Unsurprisingly, asylum superintendents and the Commissioners in Lunacy referred specifically in their reports to the pressure placed by the Irish on what rapidly became severely overcrowded institutions. The need to expand facilities caused anxiety and sometimes contention between the asylum superintendents and Committees of Visitors and the Lunacy Commissioners. One response – and the Lancashire asylums were by no means alone in implementing this kind of approach – was to expand resources through the addition of annexes and new buildings. Rainhill Asylum, for example, built originally to accommodate around 400 patients, opened an annex for an additional 1,000 patients in 1886. Prestwich Asylum, near Manchester, catered originally for 450–500 patients; by 1883 it accommodated 1,294. In 1873 a fourth county asylum opened at Whittingham in Preston to accommodate 1,100 patients. Within a decade a 674-bed annex had been added that was generally used to house 'chronic' incurable patients.[25] A fifth asylum opened at Winwick in Warrington in 1902 to house chronic patients. The Commissioners in Lunacy, pessimistic about the expansion of the

Lancashire asylum system, speculated at this point that 'The relief when it comes will be but temporary' and even before Winwick took in its first patients, the Asylum Board was searching for a suitable site for a sixth Lancashire asylum.[26]

As soon as a new facility opened – asylum or annex – it was quickly filled. The supply of pauper lunatics was seemingly inexhaustible, and time after time it was reported that many of those seeking admission were Irish. Overcrowding was a constant issue; in 1854 the Committee of Visitors at Lancaster commented on the pressure placed on the institution by the port of Liverpool and its supply of 'vagrant Lunatics' who were filling the asylum with chronic and incurable patients, marking the institution out not as a place of cure but one of containment.[27] This taxed asylum superintendents keen to retain a semblance of moral management and order within their institutions.[28] The traits of Irish patients exacerbated problems further. It was observed how demanding Irish patients could be in terms of the need for a particular kind of management and discipline. Irish patients were often admitted in poor physical health and patients from Liverpool in general were described as being 'much more seriously shattered in bodily health and condition, from poverty, dissipation, and other noxious influences incidental to large towns'.[29] These patients required intensive care and management.[30]

As the majority of Irish patients were Roman Catholics, they created additional management problems and further drained financial resources, as Visiting Committees and medical superintendents were required to attend to their religious needs. Throughout the 1860s the Lunacy Commissioners criticised Lancashire's asylums for failing to provide chaplains and chapels for the large numbers of Roman Catholic patients. As late as 1888, the Commissioners observed that Lancaster still lacked proper facilities, though by 1889 Prestwich Asylum's largely Irish, Roman Catholic patients were able to attend mass in the 'old hall', but there was no separate Catholic chapel. Asylum superintendents were also obliged to negotiate the cultural beliefs of Irish Roman Catholics and overcome the 'objections' of friends of Irish patients to post-mortem examinations.[31]

Aside from public asylum provision, huge numbers of Irish patients were accommodated in neighbouring private institutions, particularly Haydock Lodge, which in 1846 was licensed to take 400 pauper patients.[32] When new county facilities were built Haydock Lodge was emptied out, as in 1851 when Rainhill Asylum opened, but as the county asylums filled up again the numbers in private asylums soared. In 1868 it was noted that 160 'county patients' were resident in Haydock

Lodge and other private asylums.[33] Patients were also placed in local workhouses though attitudes towards the practice changed depending on the pressure on asylums and cost. After 1842, Poor Law officials were required to transfer pauper lunatics from workhouses to asylums within 14 days. Nonetheless, there was considerable pressure from the local Poor Law authorities to retain less disruptive patients in workhouses to save on expenditure. They also suggested that this would reduce pressure on the asylums. In the case of the Lancashire asylums this was a valid argument, yet in 1859 the asylum superintendents and Visiting Committees insisted – despite the severe shortage of space in many years – that the asylum was the proper place to treat insanity, even in cases where there was little hope of cure, and resisted these demands. The Lunacy Commissioners also disapproved of Boards of Guardians adopting 'the very objectionable course of erecting at the workhouses special and extensive wards for their insane poor'.[34] Their arguments took a number of forms, but emphasised the need for early transfer to asylums to maximise the chance of affecting a cure, and pointed out that even bad cases and incurables would benefit, and potentially improve, under the specialised facilities, regime and expertise found in asylums. In 1856, after visiting the workhouse wards recently established by the Select Vestry of Liverpool for idiots and incurables, Rainhill's Medical Superintendent expressed concern that the Guardians intended to retain those pronounced incurable by the Parish Medical Officers and criticised them for not moving patients rapidly enough to the asylum.[35]

The situation became increasingly difficult, as asylum overcrowding worsened and expenditure on pauper lunacy placed a huge burden on the rates, particularly the cost of maintaining paupers in asylums. In 1870 the Liverpool Guardians spent £14,000 on lunacy care, and substantial portions were allocated to Irish patients.[36] In the late 1860s the Commissioners in Lunacy and the asylum superintendents were forced to compromise their position on accommodating patients in workhouses, and even instituted patient exchanges between workhouses and asylums.[37] By the 1880s, asylums were packed with what seemed to be an endless stream of chronically insane patients. Their Visiting Committees and medical superintendents expressed increased resentment about transfers back to the asylum from workhouses, complaining of the large numbers of workhouse lunatics, suffering from 'chronic insanity in its most hopeless forms', deposited in their institutions when new facilities were made available.[38] Irish patients contributed significantly to these problems: in 1894 Toxteth Union, one of Liverpool's three Poor Law Unions, spent £16,000 on Irish cases of pauper lunacy alone.[39]

Migration and susceptibility

Despite presenting the asylum and Poor Law administrators with huge management problems and soaring costs, responses to Irish lunacy could be pragmatic and even sympathetic. Lunacy was noted in the asylums' annual reports to be a problem that particularly afflicted large manufacturing towns and seaports, which placed abnormally heavy strains on groups already living at the margins of poverty, anxious, dislocated, isolated and prone to vice and intemperance. In Rainhill's inaugural year Thomas Eccleston, the institution's Surgeon-Superintendent, reported that nearly one-fifth of admissions were Irish and one-eighth Welsh 'and probably these proportions, though in themselves large, are smaller than might have been expected, considering our neighbourhood to Liverpool'.[40] In his report for 1868, Dr Thomas Lawes Rogers at Rainhill reported on the massive increase in lunatics in the county proportional to the population over the previous decade, relating this to the 'exceptional history' of the manufacturing districts 'commencing with great prosperity and full work, inviting a large influx of population, followed by the cotton famine and all the suffering and want which it entailed'.[41] Liverpool, as a major port and manufacturing city, was identified as an additional threat to the mental state of migrants, particularly the Irish. In 1874, Dr Ley, Medical Superintendent at Prestwich Asylum, commented that 'The abundance of work and the high rate of wages obtainable in Lancashire, attract to it a never failing stream of immigrants from Ireland and elsewhere, many of whom failing in the race of life, break down and find their way into our Asylums'.[42]

Many of the Irish arriving in Liverpool planned to use the port as a stepping stone for onward migration to America; some had their entire journey prepaid and on reaching Liverpool transferred directly onto vessels heading to America. Others failed to make this journey and significant numbers of the disappointed, the failures, made their way into asylums. In 1854 John Cleaton, Medical Superintendent at Rainhill until 1858, commented on the very large number of Irish admissions in that year (42 out of 145) and noted that '[i]n three instances insanity seemed to have been produced by the excitement and embarrassment incidental to contemplated emigration to America, and in each of these cases the patients were Irish peasants'.[43] Two years later, in 1856, a year when admissions had to be restricted to just over 90 due to lack of space, Dr Rogers referred to the fact that one-third of admissions were Irish:

this will not be so much a matter of surprise, when the large extent to which the Irish element prevails in the population of Liverpool is borne in mind, and when it is also recollected that it is the port through which more of the emigrants from Ireland, to America and Australia, have to take their departure, and through which many of these poor people, who have been crushed by disappointment in a foreign land, seek to return to their native homes.[44]

In a no-win situation, those who did manage to migrate were also depicted as vulnerable and the pressure of Irish migrants from Ireland was exacerbated by the intake into Lancashire's workhouses and asylums of migrants returned from America. This provoked considerable resentment on the part of the Poor Law and asylum officials, angry at the return of impoverished and 'insane' Irish to Liverpool, as their port of departure, rather than their native country. Numerically, their impact was negligible, but they became symbolic of Liverpool's frustration at having to continually manage a flow of Irish paupers and lunatics. The cost to the poor rates of paying for Irish migrants selecting to be 'passed' back to Ireland at Liverpool's expense had long been resented.[45] One specific instance, which attracted a great deal of publicity in the local press, was the removal of paupers from America in September 1858. Of the 108 returned, 90 were Irish and 17 were listed as lunatics and epileptics. Many of these were subsequently returned to Ireland at the county's cost, though 15 remained in the workhouse.[46] It was noted that several of this group, far from being recent migrants, had resided in America for many years.

Partly as a result of their migratory experiences – the uncontrolled nature of the migration, the poor conditions on board steamships and the absence of medical inspections – Irish patients tended to be admitted to asylums in poor physical health, which placed an extra burden of care on asylum staff. In 1854, 76 of the 122 patients admitted to Rainhill were described as 'much reduced and exhausted' and many of these were Irish.[47] On admission, Irish patients were frequently described as 'thin', 'weak' and 'deteriorated' in bodily health, the consequence of poverty, unemployment and ill health. Admitted in July 1866, Michael Cunningham was in a 'very thin and reduced' state, with a feeble pulse, circulation and respiration. He died eight days after admission.[48] Such patients required extra nursing care, medicines and increased rations. Many Irish patients who refused to eat were force-fed, confounding the emphasis placed on diet and regular meals under moral management.[49] Numerous weakened patients died in the institution after several months of severe illness and many were bedridden for long periods.

Despite the pressure on the institutions, anti-Irish sentiment seeped only rarely into the asylum case notes, as in 1853 when John Birmingham was admitted to Rainhill suffering from mania. Though it was noted that he could read 'pretty well' and his mental breakdown was linked to his poverty, he was also described as 'a very stupid looking, raw Irish youth, with very little mind, and full of mischief'.[50] Yet in the same year another Irish patient was noted to be 'very hard-working and very clever at laying down drains' and it was implied that his mental breakdown resulted from trailing around the country seeking work and losing contact with his family. The illness of an Irish shoemaker, described as 'depressed and care worn', was related to 'poverty and distress', his failure to maintain his family and anxiety about the payment of rent.[51] Many of these characteristics were of course shared by other asylum patients, many of whom were migrants from other countries or regions of England, liable to the vagaries of poverty, unemployment, ill health and isolation, but Irish patients appear to have been particularly isolated and particularly liable to slip into dire circumstances.

Any sympathy – or rather pragmatism – felt by the Poor Law authorities and asylum managers towards the plight of Irish migrants was tempered by the sense that the Irish had a particular propensity towards institutionalisation and adeptness in accessing poor relief, as well as a clear understanding of their entitlements under the English Poor Law. In 1851 the Liverpool Select Vestry, still struggling with the extreme circumstances caused by Irish Famine migration into the city, commented that among the Irish 'Their object is professedly, to find work; but in reality to beg... Liverpool, in short, from its proximity to Ireland, is made the Pass House for all England'.[52] Those who failed to seek or find work became reliant on the poor rate, and many of these subsequently entered the asylum system. Many Irish patients were transferred from other institutions, mainly workhouses, but also prisons and gaols – the Irish were overrepresented in Lancashire crime statistics. According to the Reception Orders of 1873, of the 85 Irish patients admitted to Rainhill Asylum, 61 were transferred from workhouses, including their lunacy wards.

The relatively frequent – and oftentimes rapid – progress of Irish paupers from the workhouse or other institutions to the asylum is also evidenced in notebooks compiled after 1866 by the Lancashire County Council to establish settlement and the place of birth of asylum patients. They also provide evidence in some cases of complex migratory patterns

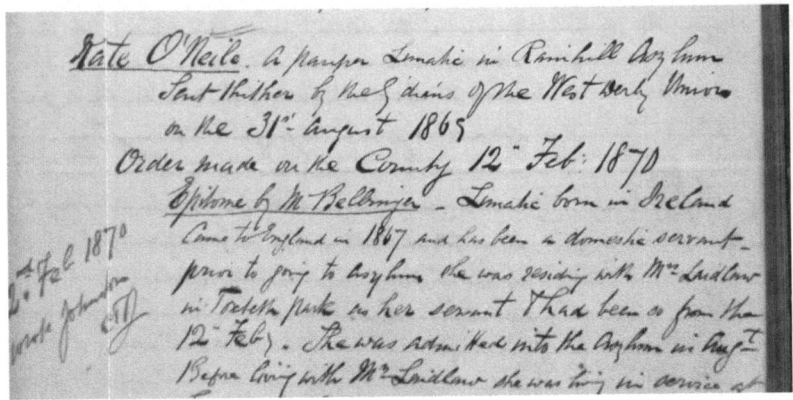

Figure 2.2   Note on Kate O'Neile in Lancashire County Council notebook, August 1869 (Courtesy of the Lancashire Archives)

In August 1869, Kate O'Neile was admitted to Rainhill Asylum on the order of the West Derby Union, suffering from mania. The County Council notebook explained how, aged 24, she had come from Ireland to England in 1867 and had worked as a domestic servant with Mrs Ludlow in Toxteth Park, and prior to that in Southport. After four years at Rainhill she was removed to Whittingham Asylum, unimproved.

Source: Liverpool Record Office M614 RAI/18/5, Rainhill Asylum, Female Casebook, 1865–1870, 327; Lancashire Archives, QAM 4/2, 9.

and insight into the fortunes of Irish patients prior to their confinement (see Figure 2.2).[53] In April 1871, for example, James Minney, aged 35, was transferred to Rainhill by the Liverpool Select Vestry. Born in County Cavan, he was sent from Knutsford Gaol to Liverpool and 'was admitted at once to the workhouse and taken thence to the asylum – had no fixed place of abode – was a tramp'. Though Minney was reported to be silent, moody and melancholy, the asylum casebook was ambiguous about his mental condition, and two months later he escaped 'relieved'.[54] Michael Meaney was sent by the Parish of Liverpool to Rainhill on 12 December 1879. He had arrived by ship from Quebec on 2 December and was admitted to Liverpool workhouse on the same day. Meaney was reported to be originally from Waterford in Ireland and the cause of his insanity was given as 'drink'. The casebook entry noted, 'Says he has been in the army for a number of years left it a few years back & has been in america'. Meaney remained in Rainhill for five years before being removed to Whittingham Asylum, the end station for many Irish patients who languished there until their deaths.[55]

Managing the degenerate Irish

The number of lunatics in Lancashire continued to grow during the late nineteenth century, and in 1901 the Commissioners in Lunacy reported that the asylum population of the country had quadrupled over the last 40 years.[56] Although by the late nineteenth century the number of Irish patients had fallen as a proportion of admissions, they were still perceived as making a massive contribution to the overwhelming pressure on asylums. In 1868 Dr Rogers noted that half of the patients resident in Rainhill in that year were English, two-fifths Irish, and the remainder originated from a variety of other nations.[57] During this later period there is evidence of a hardening of attitudes among asylum officials towards Irish patients, and in 1884 the Annual Report of Prestwich Asylum referred to the continuous problem of immigration from Ireland: 'Many of these people are persons of originally defective mental organization, who are easily upset by the hardships and worries of their new life'.[58] Asylum superintendents and the Commissioners in Lunacy continued to emphasise the problems Irish patients presented in terms of their numerical impact and the fact that they seemed to take root in the asylums to a greater extent than other patients. In 1870 Rogers referred to the

steadily increasing accumulation of Irish patients...Although the number of English admitted every year considerably exceeds that of Irish, the latter have steadily increased year by year in the residuum of incurables, until at the present time they outnumber the former, and comprise nearly one half of the whole number of patients in the Asylum.[59]

This is borne out by statistics on the length of time Rainhill patients spent in the asylum; the Irish were particularly evident among patients remaining in the asylum for over ten years (Table 2.3).

Table 2.3 Rainhill Asylum: Length of stay of Irish and non-Irish admissions, 1856–1906

Years	Irish		Non-Irish	
	Male (%)	Female (%)	Male (%)	Female (%)
≤ 1	88 (38.9)	96 (33.4)	314 (50.8)	193 (43.8)
1+–2	25 (11.1)	28 (9.8)	91 (14.7)	58 (13.1)
2+–5	27 (12.0)	47 (16.4)	80 (12.9)	62 (14.1)
5+–10	18 (8.0)	33 (11.5)	53 (8.6)	50 (11.3)
10+–20	26 (11.5)	41 (14.3)	50 (8.1)	34 (7.7)
20+–35	10 (4.4)	14 (4.9)	17 (2.8)	19 (4.3)
35+–55	1 (0.4)	4 (1.4)	2 (0.3)	7 (1.6)

Source: Liverpool Record Office M614 RAI/5-14, Admission Registers, Rainhill Asylum, 1851–1906 (database).

The social isolation of Irish migrants, which had an impact on their susceptibility to institutionalisation – in asylums, workhouses and prisons – also contributed to the length of time they spent in the asylum. After the Famine period, Irish migrants were less likely to travel in family groups, and large numbers of Irish patients were noted to be single; in 1866, 50 per cent of Irish male admissions were listed as single compared to 40 per cent of non-Irish male admissions. Reception Orders frequently noted 'nearest relative unknown', 'no friends', or that the patient's closest relatives were in Ireland.[60] In 1896, 35.6 per cent of Irish patients were still described in these terms, a decrease from 46.8 per cent in 1873 and 45 per cent in 1856.[61] Well beyond the Famine period, the Liverpool Select Vestry reported on the durable phenomenon of the 'tramping Irish' moving across the country from vagrant shed to vagrant shed, begging and committing crimes.[62] Many Irish asylum patients appear to have been excluded from the Irish community support networks that were established in towns and cities such as Liverpool.[63]

Whittingham Asylum was opened in 1873 to provide 'such supervision as is given in workhouses' to long-term, chronic patients; it was frequently used to manage Irish patients when hopes for recovery or removal had disappeared.[64] The proportion of Irish patients was falling in all of Lancashire's asylums by the last quarter of the century, but still in 1886 one-quarter of Whittingham's patient population was Irish-born – 411 out of 1,679 inmates.[65] Owen Hagan, a 48-year-old Irish widower, a fishmonger by trade, was moved to Rainhill on 13 March 1873, after being admitted to the lunatic wards at Liverpool Workhouse on 8 March 1873. The Reception Order noted that his nearest relative was unknown. He was chargeable to Liverpool Vestry and it was recorded that he had travelled from County Louth in Ireland two days prior to his admission to the workhouse. He was diagnosed with mania, was disruptive, tore up his bedclothes and suffered from delusions. On 9 September he was removed to Whittingham Asylum, unimproved.[66] Hagan was typical of many Irish patients who failed to show signs of recovery and, lacking family and friends to remove them, were transferred to Whittingham.

That the Irish were a persistent and a resented burden on asylums was an issue picked up by the local press, and in 1870 the *Preston Chronicle and Lancashire Advertiser* opined:

> many persons now in our asylums are 'importations' – people belonging to other countries, principally Ireland, who are brought to Liverpool, &c. then deserted and then picked up, placed under the

requisite surveillance, and kept at the expense of the ratepayers. If there had not been such an influx of this class, extra asylum accommodation for Lancashire would not, we feel confident, have been required yet. A much better system of supervision, by way of preventing importations of the 'finest pisantry' into Lancashire, than we have now, is required. We have plenty of insane people in the county without being put into trouble of keeping any of 'Ould Ireland's' demented children.[67]

In the late nineteenth century, Irish patients were increasingly associated with the taints of degeneration and an inherited susceptibility to mental illness and were reckoned to be particularly difficult to manage. 'Mania', marked by disruptive behaviour, accounted for just over half of annual 'Irish' admissions, while among non-Irish patients the recorded incidence was significantly lower, around 20 per cent. For example, Julia Ring, an unmarried, 40-year-old Roman Catholic Irish servant, who was described as a 'wanderer', was admitted to Rainhill Asylum on 27 March 1869. She was diagnosed with 'mania' and was noted to be disruptive, putting her dress over her head and talking 'with the most disgusting language' incessantly. She stripped herself, never kept herself tidy and was repeatedly described as 'very quarrelsome. Threatens other patients'. Two days after admission she attacked one of the nurses using considerable violence 'for a fancied wrong' and was treated with 'shower baths for excitement'. She became a long-term patient and in May 1873 was removed to Whittingham, shortly after it opened.[68]

In 1883 the Commissioners in Lunacy Report commented that Rainhill had in its care 'an excessive proportion of bad cases; many of them natives of Ireland and turbulent in disposition'.[69] Irish patients were associated with intemperance, low morality ('bad Irish character'), criminality and prostitution, as well as defective mental organisation, which also meant that they were more likely to fall into the category of 'incurable'. Alcohol and intemperance was frequently assigned as a cause of insanity, particularly among male Irish patients; in 1874 it was linked to 20 per cent of Irish male admissions to Rainhill compared to 12 per cent among non-Irish men. The Irish were, the annual reports concluded, 'failures in the race of life'. Despite these associations, there were glimmers of responsiveness to the plight of Irish patients, expressions of concern about relieving their condition as much as possible, and recognition that circumstance and environment had a role to play in triggering mental decline. Dr Rogers, Medical Superintendant at Rainhill Asylum, was especially responsive to the plight and poverty of

Irish patients and entertained some hope for their recovery. He wrote in 1866 that

> although the character of a large proportion of the patients in the Asylum, being drawn from the Irish quarters of Liverpool, is intrinsically bad and their mental condition such as to afford no hope whatever of ultimate recovery; yet by the closest attention to minute details and the willing and active exertions on the part of the officers and attendants a comparatively fair amount of order and propriety of behaviour has been established, even amongst this class.[70]

The association of large manufacturing centres with an increased likelihood of mental disease resonated particularly with Irish patients. The fact that many Irish migrants were from rural areas, and thus were depicted as 'peasants', alerted social commentators and the local press as well as asylum superintendents to their increased vulnerability when they relocated to a port city of Liverpool's size and dynamism, with its associated evils. The Irish were not only perpetrators of urban disorder, dirt, disease and vice, but also its victims.

Nowhere was this more clearly expressed than in connection with concerns about the rise of general paralysis in asylums more generally and among Irish inmates in particular. General paralysis was associated with bad living and habits, intemperance, reversal of fortune or overstrain of mental energies, and believed to be concentrated in areas where an 'active spirit of commercial enterprise' prevailed.[71] It was regarded as being generally incurable and contributed to high death rates.[72] Such patients clogged up asylums as they were rarely, if ever, discharged, and they were perceived as being very difficult to manage. Already in the 1850s, annual reports and case books referred to high levels of general paralysis and it was a common diagnosis in male Irish patients. Michael Mulloy, a married Irish labourer, was admitted to Rainhill for a second time in July 1853. He had been travelling around picking up work in Lancashire – Manchester, Bolton and Wigan – 'without any of his friends knowing his movements' and eventually he returned to his wife who became frightened by 'his strange behaviour'. By June 1854, the asylum medical officer noted that he had 'symptoms of advanced General Paralysis' and he was 'very unsteady on his legs'. Nearly four years later, he was confined to bed following the deterioration in his condition. He was destructive, requiring 'strong clothing and an occasional sedative', and died in June 1858.[73] Patients like Michael acted as a drag on recovery and discharge rates and they were increasing in number.[74] Many of these cases of a 'hopeless character', 'which those conversant with insanity know well to be an incurable and fatal disease', had been transferred from the workhouses; '*as long as this system continues, so long will our death-rate be high, and our recoveries proportionately lessened*; besides which, the tranquillity of our wards is considerably interfered with'.[75] Arthur Burns, a married smith, was typical of such cases. He was admitted to Rainhill in January 1866 and diagnosed as suffering from 'mania with general paralysis'. He was 'destructive and dirty on several occasions' and was confined to bed in a single room. The medical officer noted that it was 'doubtful whether he will ever leave it' and he was 'going the general way of paralytics'. He died in September of that year.[76]

By the 1880s it was commented that the character of general paralysis had changed significantly from being linked to younger male patients to one associated with 'older, broken down, demented, filthy, voracious creatures', who also survived much longer 'and may be seen as long as even six or seven years in our wards before the closing scene occurs'. The burden shifted from managing 'furious excitement' and 'extraordinary violence' to caring for the long-term sick, paralysed, 'degraded', with 'no mind left'.[77] It had also become more prevalent among women, some of whom were prostitutes like Matilda Fox, a 22-year-old single woman admitted to Rainhill in November in 1906. She was a difficult patient and in August it was reported that '[t]his imbecile wants too much looking after for the kitchen and has been sent back to her ward'. She eventually died in July 1911, the cause of death being recorded as cerebral syphilis.[78] Like many other chronic patients, Matilda could not contribute to the asylum workforce, thus confounding the implementation of another key aspect of moral management – regular labour in the wards, laundry, kitchen or grounds. In 1888 Prestwich's annual report observed that while general paralysis was comparatively unknown in Ireland, Scotland and in rural populations, it was prevalent in Lancashire, Middlesex and other urban centres, where conditions:

> are such as to exhaust nervous vitality and predispose to structural degenerations. The Irish peasant, in his native country, has a marked immunity from these fatal forms of brain disorders, but when transplanted into centres of labour and activity... he is often apt to break down and acquire a form of mental disease, progressive in its nature, and little susceptible of cure.[79]

Concerns about the degenerate Irish took place against the backdrop of growing interest in the increase in insanity in Ireland – as well as among Irish migrants – which was claimed to be higher than elsewhere in the British Empire. Daniel Hack Tuke, referring to rates of certification of lunatics and idiots in Irish asylums between 1875 and 1893, noted a rate of increase of 60 per cent compared with 22 per cent in England and Wales. He attributed this to many factors, including migration of the strongest, which left those behind in Ireland more liable to lunacy, as well as the return of insane emigrants, poverty, loss of land, heredity and intermarriage.[80] In a response to his article, Dr Thomas Clouston, Medical Superintendent of the Royal Edinburgh Asylum, commented

> It is well known that when a primitive people is subjected to sudden change in its surroundings, and has suddenly to adapt itself to new social conditions and environments, ... it is liable to striking consequences, such as extinction, degeneration, or loss of mental force.[81]

In making this statement, Clouston blended environment, circumstance, degenerative traits and race, something which recurs time and again in discussions of Irish insanity and the high rates of admissions of Irish patients into the Lancashire asylums.

Conclusion

While asylum medical superintendents attempted to track the number of Irish admissions and comprehend the scale of the 'Irish problem', they acknowledged that figures for Irish admissions were underestimates. The statistics did not account for patients taken from established Irish communities and who were born in England or elsewhere – individuals Rogers described as 'essentially Irish in everything but their accidental birthplace', and who were occasionally noted down as born in England of Irish parents in the case notes.[82] Even with these limitations, which reduce rather than exaggerate the proportion of Irish patients in institutions, the figures suggest that throughout the second half of the nineteenth century, the Lancashire Poor Law and asylums functioned as arms of the Irish Poor Law and asylum system. The presence of large numbers of Irish patients contributed significantly to the various challenges medical superintendents encountered when endeavouring to manage patients in the ever-expanding and overcrowded county asylums. While counties and parishes throughout England struggled to cope with pauper lunacy, the movement of Irish migrants into the Lancashire system placed additional pressures on resources and added to demands for the expansion of the county asylums time and again. Many Irish patients remained in the asylum for decades becoming part of the long-term 'chronic' patient population that 'silted-up' late nineteenth-century asylums and significantly swelled the tax burden of pauper lunacy, placing additional pressures on already overworked staff operating a 'moral management' regime that valued order and regularity. Responses to Irish patients and their management were also linked to the impact the Irish had on the Lancashire Poor Law system more generally, public health concerns and religious tensions, and the Irish were portrayed as bringing disease, disorder, violence and sectarian tensions to Lancashire and its institutions.

According to asylum records, the 'defining features' of Irish lunacy and Irish admissions – mania, intemperance, criminality and unruly behaviour – increased the challenges of managing this group of patients. Irish patients arriving in the asylums, many transferred from work-houses and prisons, were in need of extensive nursing care; they were debilitated by poverty and hardship and suffered from conditions, notably general paralysis, which took a heavy physical toll and was associated with disruptive and delusional behaviour and long-term institutionalisation concluding in death. Patients, alone and without relatives or friends willing to remove them, died within the asylums, often following transfer to Whittingham. In addition to being poorer, weaker, sicker and more isolated, Irish patients were also perceived to be more vulnerable to the stresses of city life. In the later nineteenth century, the susceptibility of Irish patients to mental illness was linked to ideas of degeneration as well as to their cultural habits and behaviour. The Irish race possessed a 'defective' mental constitution, as evidenced by the high rates of asylum institutionalisation elsewhere. The experience of migration posed serious challenges, and experience of city life in Liverpool was an even greater threat; here the already enfeebled Irish mental constitution was further undermined by the poor living conditions and moral turpitude associated with cities and industrial life, ethnicity meeting environmental challenges in a particularly disastrous form.

Notes

1. For example, see John W. Fox, 'Irish Immigrants, Pauperism, and Insanity in 1854 Massachusetts', *Social Science History*, 15 (1991), 315–36; Angela McCarthy, 'Ethnicity, Migration and the Lunatic Asylum in Early

Twentieth-Century Auckland, New Zealand', *Social History of Medicine*, 21 (2008), 47–65; David Wright and Tom Themeles, 'Migration, Madness, and the Celtic Fringe: A Comparison of Irish and Scottish Admissions to Four Canadian Mental Hospitals, c.1841–91', in Angela McCarthy and Catharine Coleborne (eds), *Migration, Ethnicity and Mental Health: International Perspectives, 1840–2010* (New York and London: Routledge, 2012), 39–54; Stephen Garton, *Medicine and Madness: A Social History of Insanity in New South Wales, 1880–1940* (Kensington: New South Wales University Press, 1988).

2. Elizabeth Malcolm, '"A most miserable looking object" – The Irish in English Asylums, 1851–1901: Migration, Poverty and Prejudice', in John Belchem and Klaus Tenfelde (eds), *Irish and Polish Migration in Comparative Perspective* (Essen: Klartext Verlag, 2003), 121–32; idem, 'Mental Health and Migration: The Case of the Irish, 1850s–1990s', in McCarthy and Coleborne (eds), *Migration, Ethnicity and Mental Health*, 15–38; Vishal Bhavsar and Dinesh Bhugra, 'Bethlem's Irish: Migration and Distress in Nineteenth-century London', *History of Psychiatry*, 20 (2009), 184–98; Liam Greenslade, Moss Madden and Maggie Pearson, 'From Visible to Invisible: The "Problem" of the Health of Irish People in Britain', in Lara Marks and Michael Worboys (eds), *Migrants, Minorities and Health: Historical and Contemporary Studies* (London and New York: Routledge, 1997), 147–78.

3. Oliver MacDonagh, *The Union and its Aftermath* (London: Allen and Unwin, 1977); Stephen Howe, *Ireland and Empire: Colonial Legacies in Irish History and Culture* (Oxford: Oxford University Press, 2000); Terrence McDonough (ed.), *Was Ireland a Colony?: Economics, Politics, and Culture in Nineteenth-Century Ireland* (Dublin: Irish Academic Press, 2005).

4. See Alison Bashford's chapter in this volume.

5. Anne Digby, 'Moral Treatment at the Retreat, 1796–1846', in W.F. Bynum, Roy Porter and Michael Shepherd (eds), *Anatomy of Madness II. Institutions and Society* (London and New York: Tavistock, 1985), 52–72; Akihito Suzuki, 'The Politics and Ideology of Non-Restraint: The Case of the Hanwell Asylum', *Medical History*, 39 (1995), 1–17.

6. The Great Famine was caused by a potato blight, which first hit Ireland late in 1845 and returned over consecutive potato harvests. Due to a combination of heavy dependence on the potato crop among agricultural labourers, limited industrial development and a heavily indebted and sometimes absent landlord class, the impact on the economy was devastating. See Mary E. Daly, *The Famine in Ireland* (Dundalk: Dundalgan Press, 1986); James S. Donnelly Jr., *The Irish Potato Famine* (London: Sutton, 2001); Peter Gray, *Famine, Land and Politics. The British Government and Irish Society, 1843–50* (Dublin: Irish Academic Press, 1998); Cormac Ó Gráda, *The Great Irish Famine* (London: Macmillan, 1989); Cormac Ó Gráda, *Black '47 and Beyond: The Great Irish Famine in History, Economy and Memory* (Princeton, NJ: Princeton University Press, 1999).

7. David Fitzpatrick, *Irish Emigration, 1801–1921* (Irish Economic and Social History: Dublin, 1984), 5.

8. Frank Neal, *Sectarian Violence: The Liverpool Experience, 1819–1914* (Manchester: Manchester University Press, 1988), 82. See also Donald MacRaild, *Irish Migrants in Modern Britain, 1750–1922* (Houndmills: Palgrave Macmillan, 1999).

9. Neal, *Sectarian Violence*, 9.

10. See J. Papworth, 'The Irish in Liverpool, 1853–71: Family Structure and Residential Mobility' (University of Liverpool PhD thesis, 1982); David Fitzpatrick, 'Irish Emigration in the Later Nineteenth Century', *Irish Historical Studies*, 22 (1980), 126–43 and Donald M. MacRaild, *Culture, Conflict and Migration: The Irish in Victorian Cumbria* (Liverpool: Liverpool University Press, 1998), 11.

11. Neal, *Sectarian Violence*, 83; idem, 'Liverpool, the Irish Steamship Companies and the Famine Irish', *Immigrants and Minorities*, 5 (1985), 28–61.

12. By the late nineteenth century some Liverpool Irish lived in ordinary working-class districts and included substantial numbers of artisans and middle-class and professional elements. Also, in this later period, migrants moved into new settlement areas outside Liverpool: see Papworth, 'The Irish in Liverpool, 1853–71'. 82 per cent, however, were listed as unskilled manual labourers: see John Belchem, *Irish, Catholic and Scouse: The History of the Liverpool-Irish 1800–1939* (Liverpool: Liverpool University Press, 2007), 27, 39.

13. *Liverpool Mercury*, 15 January 1847.

14. Neal, *Sectarian Violence*, 87.

15. House of Commons Parliamentary Papers Online: Select Committee on Poor Removal, 1854, Minutes of Evidence, Rev. A. Campbell, Rector of Liverpool, Q.4954 (parlipapers.chadwyck.co.uk).

16. Neal, *Sectarian Violence*, 107.

17. Neal, 'Liverpool, the Irish Steamship Companies and the Famine Irish', 50.

18. The 1846 Five Year's Residence Act (9 &10 Vict., Chapter 66) made it illegal for the Poor Law authorities to remove paupers who had resided in a parish for five years thereby creating the category of the 'irremovable' poor. See Neal, *Sectarian Violence*, 87. In 1861, the period of residency was reduced to three years (24 & 25 Vict., Chapter 76).

19. Liverpool Record Office (LRO) 353 SEL 10/2, Workhouse Committee Minute Book, Brownlow Hill, 4 September 1851–30 July 1853, 14 December 1851.

20. Neal, *Sectarian Violence*, 96.

21. *The Times*, 2 April 1847.

22. House of Commons Parliamentary Papers Online: First Report of the Commissioners of Enquiry into the State of Large Towns and Populous Districts, 1844 (parlipapers.chadwyck.co.uk); House of Commons Parliamentary Papers Online: W.H. Duncan, 'On the Physical Causes of the High Rate of Mortality in Liverpool' in First Report of the Commissioners of Inquiry into the State of Large Towns and Populous Districts,1844, 29 (parlipapers. chadwyck.co.uk).

23. *Daily Post*, 14 June 1871; House of Commons Parliamentary Papers Online: First Report of the Local Government Board, 1872, 40 (parlipapers.chadwyck.co.uk).

24. LRO H352.4/HEA, Medical Officer of Health Reports, W.S. Trench, *Report on the Health of Liverpool during the Year 1866* (Liverpool, 1867), 21.

25. Whittingham added an acute ward for recent cases of insanity in 1899 to 'favourably influence the recovery rate'. House of Commons Parliamentary Papers Online: Fifty-Fifth Annual Report of the Commissioners in Lunacy, 1901, 307 (parlipapers.chadwyck.co.uk).

26. Ibid., 28.

27. Wellcome Library (WL), Annual Report, Lancaster Asylum 1854, Report of Committee of Visitors, 9.
28. Suzuki, 'The Politics and Ideology of Non-Restraint'; Andrew Scull, *The Most Solitary of Afflictions: Madness and Society in Britain, 1700–1900* (New Haven, NJ and London: Yale University Press, 1993), 370–4.
29. WL, Annual Report, Rainhill Asylum 1856, Report of Medical Officer and Superintendent, 94–5.
30. See Catherine Cox, Hilary Marland and Sarah York, 'Emaciated, Exhausted and Excited: The Bodies and Minds of the Irish in Nineteenth-Century Lancashire Asylums', *Journal of Social History*, 46:2 (2012), 1–26.
31. House of Commons Parliamentary Papers Online: Thirtieth Report of the Commissioners in Lunacy, 1876, 189; Forty-Second Report of the Commissioners in Lunacy, 1888, 199; Forty-Third Report of the Commissioners in Lunacy, 1889, 225; quote in Forty-First Report of the Commissioners in Lunacy, 1887, 216 (parlipapers.chadwyck.co.uk).
32. W. Ll. Parry-Jones, *The Trade in Lunacy: A Study of Private Madhouses in England in the Eighteenth and Nineteenth Centuries* (London: Routledge, 1972), 58.
33. LRO M614 RAI/30/1, Committee of Visitors Minute Books, Minutes 1851–62, Adjourned Meeting, 14 April 1851; M614 RAI/40/2/1, Annual Report, Rainhill Asylum 1868, Report of Medical Superintendent, 103.
34. House of Commons Parliamentary Papers Online: Thirteenth Report of the Commissioners in Lunacy,1859, 17 (parlipapers.chadwyck.co.uk).
35. LRO M614 RAI/40/2/26, Annual Report, Rainhill Asylum 1856, Committee of Visitors, 58–9.
36. *Daily Post*, 22 July 1870.
37. Attempts to regulate the transfer of patients from workhouses to asylums were introduced under the 1862 Lunacy Acts Amendment Act: see Elaine Murphy, 'The Lunacy Commissioners and the East London Guardians, 1845–1867', *Medical History*, 46 (2002), 512.
38. LRO M614 RAI/40/2/6, Annual Report, Prestwich Asylum 1888, Report of Medical Superintendent, 69.
39. LRO 353/1/1, Toxteth Park Union, Minute Books, 1894–96, 11 October 1894.
40. WL, Annual Report, Rainhill Asylum 1851, Report of Medical Superintendent, 10.
41. LRO M614 RAI/40/2/1, Annual Report, Rainhill Asylum 1868, Report of Medical Superintendent, 103–4.
42. LRO M614 RAI/40/2/2, Annual Report, Prestwich Asylum 1874, 69.
43. WL, Annual Report, Rainhill Asylum 1854, Report of Medical Superintendent, 81.
44. LRO M614 RAI/40/2/26, Annual Report, Rainhill Asylum 1856, Report of the Medical Resident Officer and Superintendent, 85.
45. Neal, 'Liverpool, the Irish Steamship Companies and the Famine Irish', 48–51.
46. *Daily Post*, 15 September 1858; *Liverpool Mercury*, 12 November 1858.
47. WL, Annual Report, Rainhill Asylum 1854, Report of Superintendent, 85.
48. LRO M614 RAI/11/4, Rainhill Asylum, Male Casebook, 1865–1870, no. 2536, Michael Cunningham, 7 July 1866.
49. For examples, see LRO M614 RAI/11/4, Rainhill Asylum, Male Casebook, 1865–1870, William McGlone, 4 January 1866; M614 RAI/11/5, Rainhill

Asylum, Male Casebook, 1870–1873, Michael Manley, 11 October 1871; M614 RAI/8/6, Rainhill Asylum, Female Casebook, 1870–1873, Maria Walsh, 10 July 1871.

50. LRO M614 RAI/11/1, Rainhill Asylum, Male Casebook, 1853–1857, no. 748, John Birmingham, 30 July 1853.
51. Ibid., no. 743, Michael Mulloy, 20 July 1853; no. 747, Peter Edmonds, 30 July 1853.
52. LRO 353/SEL/10/2, Workhouse Committee Minute Book, Brownlow Hill, 4 September 1851–30 July 1853, 4 December 1851.
53. The notebooks produced by the Lancashire County Council, General Finance Committee placed all lunatics charged to the County in two classes; the notebook compiled for class 1 patients contained names and details of those who had no settlement in England and Wales, including Irish-born patients, who were to be maintained at the cost of the county. Lancashire Archives (LA), QAM 4/1: Register of Class 1 lunatics, covering admissions 11 December 1866–31 August 1869.
54. LRO M614 RAI/11/5, Rainhill Asylum, Male Casebook, 1870–1873, James Minney, 28 April 1871, 70; LA, QAM 4/2, 43.
55. LRO M614 RAI/11/7, Rainhill Asylum, Male Casebook, 1877–1881, Michael Meaney, 12 December 1879, 212; LA, QAM 4/2, 282.
56. House of Commons Parliamentary Papers Online: Fifty-Fifth report of the Commissioners in Lunacy, 1901, 8 (parlipapers.chadwyck.co.uk).
57. LRO M614 RAI/40/2/1, Annual Report, Rainhill Asylum 1868, Report of Medical Superintendent, 104.
58. LRO M614 RAI/40/2/5, Annual Report, Prestwich Asylum 1884, Report of Medical Superintendent, 65.
59. LRO M614 RAI/40/2/1, Annual Report, Rainhill Asylum 1870, Report of Medical Superintendent, 115.
60. LRO M614 RAI/1/3-4, Reception Orders, Rainhill Asylum, Nos. 2501–2550, October 1865–April 1866.
61. LRO M614 RAI/1/1, Reception Orders, Rainhill Asylum, December 1871, Nos. 1046–1100; M614 RAI/1/3, Reception Orders, Rainhill Asylum, October 1865–April 1866, Nos. 2501–2550; M614 RAI/1/5, Reception Orders, Rainhill Asylum, December 1871–May 1874, Nos. 3501–3850; M614 RAI/1/25, Reception Orders, Rainhill Asylum, October 1895–December 1896, Nos.10451–10797.
62. David Fitzpatrick, '"A Peculiar Tramping People": The Irish in Britain, 1801–1870', in W.E. Vaughan (ed.), *New History of Ireland VI. Ireland under the Union 1: 1870–1921* (Oxford: Clarendon, 1989), 623–60.
63. For a detailed discussion, see Cox, Marland and York, 'Emaciated, Exhausted and Excited'.
64. House of Commons Parliamentary Papers Online: Thirty-Ninth Report of the Commissioners in Lunacy, 1884–85, 236 (parlipapers.chadwyck.co.uk).
65. LRO M614 RAI/40/2/5, Annual Report, Whittingham Asylum 1886, Report of the Commissioners in Lunacy, 152, 170.
66. LRO M614 RAI/1/5, Reception Orders, December 1871–May 1874, No. 3528, Owen Hagan; M614 RAI 11/5, Rainhill Asylum, Male Casebook, 1870–1873, Owen Hagan, 13 March 1873, 209; LA, QAM 4/2, 90.
67. *Preston Chronicle and Lancashire Advertiser*, 18 June 1870, 6.

68. LRO M614 RAI/8/5, Rainhill Asylum, Female Casebook, 1865–1870, Julia Ring, 27 March 1869, 305.
69. LRO M614 RAI/40/2/5, Annual Report, Rainhill Asylum 1883, Commissioners in Lunacy, 95.
70. LRO M614 RAI/40/2/31, Annual Report, Rainhill Asylum 1866, Report of the Medical Superintendent, 106.
71. WL, Annual Report, Rainhill Asylum 1854, Report of the Superintendent, 81.
72. Gayle Davis, 'The Cruel Madness of Love': Sex, Syphilis and Psychiatry in Scotland, 1880–1930 (Amsterdam and New York: Rodopi, 2008), 83–7.
73. LRO M614 RAI/11/1, Rainhill Asylum, Male Casebook, 1853–1857, no. 743 Michael Mulloy, 20 July 1853.
74. For a more extended discussion of the impact of general paralysis, see Cox, Marland and York, 'Emaciated, Exhausted and Excited'.
75. LRO M614 RAI/40/2/1, Annual Report, Prestwich Asylum 1866, Commissioners in Lunacy, 54; Annual Report, Prestwich Asylum 1870, Report of the Superintendent, 52–3, 56 (their italics).
76. LRO M614 RAI/11/4, Rainhill Asylum, Male Casebook, 1865–1870, Arthur Burns, 17 January 1866, 40.
77. LRO M614 RAI/40/2/6, Annual Report, Whittingham Asylum 1887, Report of the Medical Superintendent, 164–5.
78. LRO M614 RAI 8/25, Rainhill Asylum, Female Casebook, 1905–1906, Matilda Fox, 12 November 1906, no. 15,274, 82.
79. LRO M614 RAI/40/2/6, Annual Report, Prestwich Asylum 1888, Report of the Medical Superintendent, 69. See also Thomas More Madden, 'On the Increase of Insanity, with Suggestions for the Reform of Lunacy Laws and Practice', Dublin Journal of Medical Science, 78 (July–December, 1884), 303–14, 304–5.
80. Daniel H. Tuke, 'Increase of Insanity in Ireland', Journal of Mental Science, 40: 171 (October, 1894), 549–61, 549, 561.
81. Ibid., 561. See also Catherine Cox, Negotiating Insanity in the Southeast of Ireland, 1820–1900 (Manchester: Manchester University Press, 2012), 53–65 for a full and more nuanced take on the debate on the incidence of Irish insanity in the late nineteenth century.
82. LRO M614 RAI/40/2/1, Annual Report, Rainhill Asylum 1870, Report of the Medical Superintendent, 115.

3
Migration and Mental Illness in the British West Indies 1838–1900: The Cases of Trinidad and British Guiana

Letizia Gramaglia

The development of the asylum system in the British West Indies coincided with the aftermath of the abolition of slavery, a period of intense social transformation and diversified internal and external labour migration. The onset of a system of apprenticeship, a six-year interim phase between abolition and actual freedom set out by the English Parliament to reduce the burden of mass emancipation, made cheap plantation labour a pressing need.[1] Planters in some of the larger colonies endeavoured to import workers from neighbouring sugar territories, generating erratic yet significant migration flows in the region. From as early as 1835, increasing numbers of freed slaves moved from the most populated areas, particularly Barbados, St Kitts and Antigua, to less densely populated colonies of the British West Indies, including British Guiana, Trinidad and Jamaica. Small waves of European labourers, mainly Portuguese from Madeira, were also introduced to the region after 1835 to work on sugar plantations, and from 1838 onwards large numbers of contracted workers, from India and China, were recruited to replace slave labour. Although on a much smaller scale than major movements of population from Europe to North America over the same period, this phase of immigration into the West Indies had a lasting and decisive impact on the social, economic, political and cultural history of the region.[2]

Drawing on immigration records and the pioneering observations of three British doctors deployed to work in the lunatic asylums of Trinidad and British Guiana, this chapter explores the relationship between migration and mental illness in the two colonies during the second half of the nineteenth century. What emerges from these documents is a link between the conditions of migrant labour and the high incidence of mental illness in the colonies, with the local treatment of mental health developing within the predominant ethnic or racial taxonomy of the period. The doctors – Robert Grieve, Resident Medical Superintendent at the Guianese Asylum, his predecessor the Resident Surgeon James S. Donald, and George Seccombe at the Belmont Lunatic Asylum in Port of Spain – produced documentary representations of asylum life, and observed, to various degrees, the challenges of migration as they related to clinical practices and medical advances and initiatives. Foregrounding three key factors in the mental health of the patients that came under their care – biological, contextual and behavioural – they argued that racial propensities, poverty, substance misuse, the stress of migration and exposure to new environments were all potentially central in triggering mental illness.

Trinidad and British Guiana are important contexts where the large-scale immigration of indentured labour from India during the nineteenth century pressured colonial governments to become more proactively involved in questions of social and medical welfare. Before 1838 'planters contracted European doctors to visit their estates regularly and attend the enslaved population', but this practice declined with the abolition of slavery, causing an exodus of European physicians from the region and the shifting of medical costs from planters to the free labouring population who were now obliged to pay prohibitive fees for medical care.[3] However, a more complex scenario developed in British Guiana and Trinidad, engaged as they were in the mass importation of indentured immigrants, with planters and governments keen to protect both their reputation and their human assets. As K.O. Laurence has pointed out, the problem of providing medical assistance for the new immigrants in these two territories saw planters and local administrations debating for a long time over issues of welfare responsibility. At the time of abolition, neither colony had a centralised public health service. Georgetown and Port-of-Spain had overcrowded and deficient public hospitals, and there was no provision at all in the country districts where the labourers were deployed. In both colonies the government ultimately resorted to 'compulsion on the employers' to provide some kind of medical assistance for their workers. In British Guiana, the immigrant population was much larger and the rate of sickness among new labourers was extremely high. This led the Court of Policy, the local executive and legislative body presided over by the Governor of

the colony, to pass an ordinance in 1847 to compel planters to maintain a hospital on each estate and employ a doctor to visit the site every
48 hours.[4] A similar ordinance was passed in Trinidad in 1866, forcing
planters to provide estate hospitals and cover the costs of medical care,
on pain of the removal of immigrants from their estates. Yet, enforcement remained difficult and sporadic. The Commission of Enquiry that
visited British Guiana in 1870 to investigate the treatment of immigrants conducted a thorough review of estate hospitals and concluded
that, although the system as a whole was worthy of praise, it was still
lacking in many respects. The Commission recommended that doctors
should be paid by the government rather than the planters,[5] and by
1873 both British Guiana and Trinidad had established a system of
District Medical Officers in charge of estate hospitals funded by local
colonial governments.[6]

The second half of the nineteenth century also saw administrations
across the British Empire becoming increasingly concerned with the
mental health of their subjects as growing levels of poverty, deprivation and associated mental illnesses threatened the social order of the
colonies. Trinidad and British Guiana, which drew the highest number
of immigrants, took the lead in providing an alternative to prison detention for those individuals who displayed signs of mental illness. Lord
Harris, the Governor of Trinidad, wrote to the Secretary of State for the
Colonies to point out the shortfalls in custodial arrangements for the
insane. His dispatch, dated 21 February 1848, stressed the urgent need
for dedicated facilities as '[t]he lunatics and the idiots wander at large
about the streets, to the annoyance and disgust of all, except when at
times they become violent; then, if by chance room may be found either
in the gaol or the hospital, or the police station, they are confined. Daily,
during the last year, have I desired to commence building only a few
strong rooms, in which they might be housed, but the want of funds
has stopped me'.[7] Harris's words not only identify the lack of welfare
infrastructure and provision for the ever-increasing migrant workforce
and the mentally ill population in the West Indian colony, they also suggest the dominant rhetoric of nuisance, marginalisation and repulsion
that surrounded mental health issues.

Migration

Critically, once lunatic asylums were established in Trinidad and British
Guiana, the vast majority of people committed to the institutions were
immigrant workers. Fairly reliable immigration records were kept in the

British West Indies in the post-emancipation period, mainly due to the fact that the mass mobilisation of the labour force at this time was characterised by strong government intervention and was for the most part implemented through the expenditure of public funds.[8] In the wake of the 1834 Slavery Abolition Act, increasing numbers of people moved from the most populated areas within the region to less densely populated colonies. Indeed, until the early 1860s, internal movements of migrant workers from the Eastern Caribbean to colonies such as Trinidad and British Guiana outnumbered indentured immigration from outside the region.[9] Barbados was the main place of origin for internal migration;[10] as the oldest of the British colonies in the West Indies (along with Jamaica), the island's economy was characterised by a high-density labouring population and a limited amount of unoccupied land. At the same time, the larger and less-densely populated colonies of Trinidad and British Guiana welcomed and encouraged immigration in an effort to secure a labour force for their extensive plantations.[11] In 1836 British Guiana's chief emigration agent, Thomas Day, went to Barbados seeking to hire subagents and 'offered to pay £10 for each emigrant enlisted, a sum equal to a local worker's yearly income'.[12] While the Barbadian government repeatedly expressed opposition to mass emigration 'between 1863 and 1875, the government of British Guiana subsidized the passage of an estimated 21,000 emigrants from Barbados'.[13] There was significant migration to British Guiana and Trinidad from the other Windward islands, such as Grenada, St Vincent and St Lucia. In Trinidad, the Returns for 1846 show that more than half of the working population of the colony was made up of immigrants, the majority of which were classified as 'old islanders' from the old British West Indian islands.[14] The rapidly changing human landscapes of the Caribbean, with the explosive growth of new post-emancipation economies, thus presented social challenges even for 'internally' dislocated migrants.

Meanwhile, the two main streams of external migration originated from India and China, where local conflicts, poverty and overpopulation pushed large numbers of people to sign contracts of indenture offered to them by recruiters and to depart on the arduous journey across to the Caribbean. Already in 1806 just under 200 Chinese men had been brought to Trinidad as indentured workers on the sugar estates.[15] In British Guiana, the suggestion to look to China for a supply of free labour dated back to 1811, but it was not until January 1853 that the first shipment of 262 Chinese immigrants arrived in the colony; the mortality rate during this first voyage was 16.7 per cent.[16] Two years later immigration from China was suspended due to abuses in the enlisting process, and in 1859 voluntary contracted emigration from licensed depots replaced the old system. Shipments of labourers continued on an annual basis until 1866 and the scheme finally came to an end in 1879; by then it had brought around 18,000 Chinese immigrants to the British West Indies, 13,541 to British Guiana alone. Only a small fraction of these – 2,075 – were women.[17] This mass economic migration, combined with such a great imbalance in the sexes of migrants, was to be regarded as a contributing factor in the context of mental health in the recipient colonies.

The largest immigrant group landing on the shores of Trinidad and British Guiana after 1834 were East Indians, imported en masse to the West Indies under a state-regulated indentureship system over a period of more than 70 years. Described by many as a new system of slavery, Indian indentureship was based on the stipulation of a five-year contract that bound workers to work and live on a specific plantation. The so-called 'coolie trade' from India, which mobilised an international network of recruiters, shippers and investors, began in 1836, under the aegis of Liverpool merchant John Gladstone, father of the Prime Minister William Gladstone. When John Gladstone enquired about the possibility of transporting East Indian workers to his plantations in British Guiana, he was reassured that no 'difficulty would present itself in sending men to the West Indies, the natives being perfectly ignorant of the place they agree to go to, or the length of the voyage they are undertaking'.[18] On 5 May 1838 the first load of 437 Indian 'coolies' (405 men, 12 women and 20 children) were delivered to British Guiana; indentured immigration from East India to the West Indies continued steadily between 1845 and 1917, bringing over 145,000 Indians to Trinidad and 238,000 to Guiana.[19] Traditional push and pull factors led millions of Indians to sign a contract of indenture; poverty, unemployment and famine were widespread in the regions targeted by recruiters, while deception, kidnapping and enticement were commonly used to compel migrants to sign contracts and fulfil recruiters' migration quotas. Peter Ruhomon, a retired civil servant stationed in British Guiana, reported that 'the grossest deceptions have been practiced on [the immigrants] by the recruiters, in holding out to them, prospects of great fortunes to be made in a land flowing with milk and honey'.[20]

But the reality they faced was different. In a pamphlet published in 1840, John Scoble, an active British abolitionist, harshly denounced the conditions under which the coolies were recruited and detained before being forced on board the ships; the ill treatment and abuse of workers on the plantations; and the inadequate medical care offered to the injured or sick.[21] The majority of the labourers were single men in their twenties and thirties; they came mostly from the north and south of India, belonged to a wide range of castes and spoke a multiplicity of languages and dialects. The agriculturists among them were not accustomed to the hardship of plantation labour. Large numbers of the immigrants were physically unfit for work or completely unfamiliar with fieldwork. Many of these were destitute or were prostitutes who had no better prospect in life, or were deprived Brahmins and functionaries who had been forced to migrate due to a decline in their fortunes. The Commission of Enquiry observed that of the 30 adult immigrants taken to British Guiana on board the *Medea* in 1870, 14 were 'priests, weavers, scribes, shoemakers, beggars and so forth' and concluded that 'the immigrants on arrival find they have to do work to which they have never been accustomed; they get disheartened, and soon find their way into the estates' hospitals'.[22]

The psychological implications of the abuses and irregularities linked to the system of indentureship were severe. The strain of separation from home, combined with what David Dabydeen has described as the 'trauma of accommodating to the new environment', contributed strongly to the emergence of mental illness among the East Indian population in the West Indies.[23] The journey across the ocean was arguably one of the most traumatic ordeals faced by the indentured labourers, and it was often on board the ships that the Indians first encountered morbidity and mental illness.[24] The *Journal of a Voyage* kept by Captain Swinton and his wife on the *Salsette*, a ship transporting indentured immigrants to Trinidad in 1858, provides an invaluable insight into the miserable conditions of the passengers. Over the 108 days spent at sea, 124 out of the 324 Indian emigrants died; one-third of the victims were infants or children who died from poor nutrition. Shortly before the end of the journey Swinton noted in his *Journal*: 'Mustered the Coolies, and find only 108 men, 61 women, and 30 children under ten years of age, 2 infants, and 2 interpreters, left of the 323 or 324 we sailed from Calcutta with, and 3, I fear, will die before we can get them landed'.[25] Dr Mitchell, subsequently appointed by the colonial government to investigate the causes of mortality on board the *Salsette*, observed in his report to the Governor that for a considerable time the survivors 'remained in a depressed state, the effects of death and despondency endemic in their unforgettable experience of seasickness'.[26]

From a sociocultural point of view, the journey across the ocean had also a strong symbolic significance. As Kahl Torabully explains, the prohibition to cross the *Kala Pani*, the 'dark waters', represented a major taboo in traditional Hinduism: 'The soul of the Hindu who left the Ganges was doomed to err perpetually, as it was cut from the cycle of reincarnation. So going on the high seas was...a major symbolic act, as the *Kala Pani* was peopled by *houglis*, foul spirits and monsters'. Undertaking the voyage meant not only physical dislocation, but also the betrayal of strongly embedded personal beliefs and social tenets, with ensuing 'guilt complexes'.[27] This state of mental and physical distress resulting from the journey was often exacerbated when the migrants encountered the hostile human and natural environments of the Caribbean. On the plantations, the immigrants' life was made miserable by biased laws, arbitrary incarceration, unfair wages, scarcity of women, inadequate accommodation and other daily aggravations. In December 1869 William Des Voeux, the former stipendiary magistrate in British Guiana, wrote a detailed letter to the Secretary of State for the Colonies reporting on the ill treatment of immigrant labourers in the colony. More specifically, Des Voeux's memorandum charged immigration agents, magistrates, medical attendants on the estates and the late Governor Hincks, with acquiescing to the demands of the planters, and called for an investigation into the abuses of the indentureship system. According to him, inadequate medical facilities, summary and partial justice, scanty accommodation and daily harassment were common practice in the colony and reduced the indentured immigrants to a position 'not far removed from slavery'.[28]

The distress entailed by migration and adaptation, coupled with excessive and biased regimentation, led to the proliferation of mental health issues among the indentured populations of Trinidad and British Guiana. Significantly, mental illness became a dominant discourse in the collective imaginary of Indo-Caribbean immigrants, often finding expression in popular culture. An early example can be seen in the following verses, where the lyricist voices the labourers' frustration at the lack of freedom governing their lives:

> It drives one out of his mind,
> British Guiana drives us out of our minds.
> In Rowa there is the court house,
> In Sodi is the police station,
> In Camesma is the prison.
> It drives one crazy,
> It is British Guiana.
> The court house in Wakenaam,
> The police station in Parika,
> The prison in Georgetown,
> Drive you crazy.[29]

As Kusha Haraksingh has suggested, the indentured labour force was 'held "captive" at several different but connected levels ... In psychological terms, the prison walls were clearly discernible, for even the ordinary run of plantation life fostered a general feeling of helplessness'.[30]

Mental health provisions and organisation

As noted, it was in this climate of intense and traumatic labour relocation that the asylum system was born in the British West Indies, as part of the development of public services that slowly followed emancipation. Curative strategies and models of institutional management that had been already adopted in Britain were introduced mainly through the appointment of experienced doctors and implemented to various degrees according to the size and wealth of the colonies. Jamaica was the only territory in the region to have a specialist institution in place before the 1830s, but in the two decades following emancipation most colonies in the British West Indies established some sort of mental institution.

The Belmont Lunatic Asylum in Port of Spain opened in 1858;[31] this establishment, which survived for more than 40 years, housed an ever-increasing number of patients, rising from 48 inmates in 1860 to nearly 500 in 1899.[32] The extensive number of East Indians in Trinidad was reflected in asylum admission figures available from the early 1880s to the late 1890s; for example, in 1883, nearly 44 per cent of the 103 patients admitted were Indian immigrants.[33] In 1887, there was a small decline in the admission of East Indians; however, of the 32 Indian patients admitted during the year, 16 were women. This was disproportionately higher than the proportion of women in the population, the male to female ratio of Indian migrants in the colony being 1:3.[34]

In British Guiana, the first rudimentary institution for the mentally ill was established in June 1842, as part of Governor Henry Light's programme of social improvement.[35] Over the following decades, unsatisfactory structural and hygienic conditions resulted in several relocations, until a permanent site for the lunatic asylum was found in 1867 at Fort Canje near New Amsterdam, adjacent to the Berbice General Hospital. During the first five years, there were 195 admissions to the asylum (135 patients), who with the exception of a few Europeans and a small number described as 'mulatto', were recorded as being of African origins.[36] The demographics of the asylum changed significantly over the following decade, with immigrants rapidly becoming the largest group of patients.

In April 1876, Dr James S. Donald, Resident Surgeon at the Berbice Asylum, published an insightful article in the *Journal of Mental Science* on the incidence of mental illness in British Guiana and its relationship to ethnicity and geographic and cultural background. Donald's opening remarks read: 'Few countries, if any, afford better opportunities for the study of insanity, as exhibited among different races, than British Guiana. Here, gathered together in one asylum, are West Indians, Coolies from India, Chinese, Portuguese, and Africans; and, although the types of insanity are very similar in all, yet there are some distinctive features, worthy... of being noted'.[37] The article shows a sensitivity to issues of language and culture that was rarely present in the literature produced by his contemporaries. Donald mentions the difficulties in gathering reliable information on patients' history due to the language barrier: 'Many difficulties arise in investigating the subject, owing, principally, to an inability to converse personally with some of the patients, more especially with Chinese'.[38] Most importantly, he suggested that to compare the proportion of insane in British Guiana to that of the insane in England would be incongruous 'owing, mainly, to the fact that in many cases national peculiarity is mistaken for mental derangement'.[39] His reflections on the impact of linguistic barriers and the Eurocentric limitations of colonial psychiatry are remarkably ahead of his time, as is his emphasis on the importance and relevance of the patient's voice.

The expansive ethnic categories utilised by Donald and his contemporaries allowed for flexibility in diagnosing and categorising his patients, under a system that equated and attached economic, cultural, social and biological values to racialised groups. In certain cases, cultural practices and behaviour defined specific propensities such as intemperance in the use of alcohol in the case of Creoles and Portuguese; elsewhere, illness was attributed to economic and environmental factors, as in the case of malnutrition in relation to the East Indians. Donald noted that on admission, East Indian patients 'are almost invariably very anaemic and half-starved, owing to the insufficient nourishment which they take prior to being admitted', but also recognised that '[w]hile the percentage of admissions of East Indian immigrants is greater than that of any other nationality represented, the number of recoveries is also relatively greater. I attribute this in a great measure to the improved dietary which they receive in asylums'.[40] Conversely, his theorising in relation to Chinese patients is varyingly physiological and psychosomatic:

Among the Chinese inmates I have been struck with the frequency of epilepsy and epileptic mania, and have been equally puzzled to account for it...As a rule the characteristic stolidity and impassive-ness of the Chinese is little altered during mental aberration. The cheerless, unhappy expression of countenance gives the patient the appearance of one suffering from profound melancholia, and totally indifferent to anything around him. The number of Chinese now in the asylum is too small to warrant my giving any decided opin-ion as to what may be considered the more prominent features and nature of their mental disease. They are generally quiet docile patients, very amenable to treatment, and, except in the epileptic, violent symptoms are rare.[41]

In terms of ethnic classification of his patients, Donald's system failed to distinguish between those coming from different areas within the region, and grouped together various patients from Guiana and Barbados as 'Creoles' constituting about 25 per cent of the inmates of the Guianese asylum. He observed a prevalence of what he classi-fied as 'mania' among this group, often complicated with delusions of a religious character. Mania and dementia were also indicated as the primary affliction of the 'coolies', who accounted for over 50 per cent of the Donald's patients: 'The mania of the Coolies is generally characterised by great destructiveness and impulsiveness; consequently homicidal and suicidal propensies are of frequent occurrence among them. While, however, such cases are dangerous, they seldom last long in the acute stage, thus contrasting strongly with the form of acute mania met with in the black Creole'.[42] In colonial contexts the issue of diagnosis is one that generally raises more questions than it answers. From 1881, for example, the reports accompanying the Blue Books for British Guiana tend to classify the inmates of the Berbice Asylum under four broad categories – namely 'quiet chronic', 'maniacal', 'idiotic, paralytic and epileptic' and 'melancholy acute' – without any further specification of symptoms and signs.[43] As James Mills has pointed out, during the nineteenth century the categorisation of certain types of behaviour varied considerably among different asylums and depended very much on the judgement of individual medical officers, thus mak-ing it difficult for the reader to assign specific significance to them.[44] This was the case in the British West Indian colonies where even analo-gous classification in the colonies' official records does not infer shared semantic parameters. Thus Donald's description of 'destructiveness and impulsiveness' as well as 'homicidal and suicidal propensies' might reference behaviours which emerged simply as a form of resistance to the extremely exploitative and inhuman conditions of indenture-ship. In this sense, his pathologising of social or political opposition seems inherent to the colonial systems of control at work in British Guiana.

More extensive observations on the aetiology of insanity in different 'races' were offered by Dr Robert Grieve, appointed Medical Super-intendent of the Guianese Asylum in 1875, and a pioneer of moral management in the colony. Grieve's distaste for intellectual inertia, which he condemned as one of the highest faults of colonial life in the tropics, and the urge to overcome the isolation compelled by the asylum's geographical remoteness, prompted him to devise new means of communication with the external world. He set up a small printing office at the Berbice Asylum and, in February 1881, launched *The Asy-lum Journal*, a monthly pamphlet aimed at increasing public interest in the institution and adding to the knowledge of local medicine.[45] Grieve authored most of the *Journal* himself, and a small number of inmates were employed in making and stitching copies. His notes on mental illness included detailed case studies and articles on a broad range of topics: the aetiology of insanity in different races; the relationship of insanity with food, crime and drugs, like ganje; aspects of treatment such as work, non-restraint and specific African approaches, including prolonged seclusion and the administration of a special diet. In par-ticular, Grieve noticed the high vulnerability of Indian immigrants, in comparison to Creoles and other ethnic groups. The figures he collected showed 'a striking diversity in the proportion of inmates given to the asylum by the two greatest classes of the population here, native born and immigrants'.[46]

As Table 3.1 shows, out of the 505 admissions recorded over the first five years of Grieve's administration, only 171 were born in British Guiana, while 329 were immigrants and more than half of these were natives of India. At the same time, the large number of admissions to the asylum might have also been influenced by pecuniary considera-tions; an ordinance passed in 1864 in British Guiana established that if field workers were sent to hospital the cost of their treatment would fall on the planter. However, if they were certified insane and sent to the asy-lum, then the employer was 'relieved from any liability to payment'.[47] It seems legitimate to assume that this might have had an impact on the proportion of those classified as insane in the colony.

The publication of the 1881 census enabled Grieve to determine the exact proportion of the insane in relation to the different classes of

Table 3.1   Admissions to Berbice Asylum, 1876–1880

Natives of	1876			1877			1878			1879			1880			Total		
	M	F	T	M	F	T	M	F	T	M	F	T	M	F	T	M	F	T
British Guiana	12	21	33	18	23	41	13	14	27	16	12	28	17	15	32	76	85	171
India	22	15	37	18	18	36	27	8	35	22	12	34	35	8	43	124	61	185
Madeira	6	0	6	4	3	7	2	2	4	4	2	6	5	2	7	21	9	30
China	3	2	5	4	0	4	5	0	5	0	1	1	3	0	3	15	3	18
Africa	6	3	9	8	2	10	6	2	8	6	0	6	3	3	6	29	10	39
Barbadoes	9	3	12	4	6	10	8	3	11	5	5	10	2	3	5	28	20	48
Other W.I. islands	0	1	1	2	1	3	3	0	3	0	1	1	1	3	4	6	6	12
Europe	1	0	1	1	0	1	0	0	0	3	0	3	2	0	2	7	0	7
Unknown	1	0	1	0	0	0	2	0	2	0	2	2	0	0	0	3	2	5
Total	60	45	105	59	53	112	66	29	95	56	35	91	68	34	102	309	196	505

Source: The Asylum Journal (No. 3, 1881), 33.

the population. The number of inmates in the asylum – the only one in the colony – on the day of the census was used by Grieve as the basis of his calculation (see Table 3.2):

Table 3.2   Inmates of the Berbice Asylum according to place of origin, 1881

Class of population	No. of insane	Proportion of insane per 10,000 of population
Natives of British Guiana except aborigines	141,983	108	7.6
Natives of West India Islands	18,318	40	21.8
Natives of Madeira, &c.	6,879	15	20.3
Natives of Europe	1,617	1	6.1
Natives of North America	205	0	0.0
Natives of Africa	5,077	14	27.5
Natives of India	65,161	142	21.7
Natives of China	4,323	11	20.4
Other places or not known	897	1	11.1
Total	244,530	331	13.5

Source: The Asylum Journal (No. 13, 1882), 169.

In his analysis of the data, Grieve highlighted the striking disparity between the number of insane patients who were natives of the colony (7.6 per cent) and those who were immigrants (21.7 per cent).

In discussing the susceptibility of East Indian migrants, he identified four factors worthy of note: the possibility that immigrants would be living alone rather than in villages or groups as in the case of the Creole population; the change of circumstances brought about by migration, with the consequent separation from country and friends; the fact that mentally unstable people were more easily persuaded by the recruiters to leave their country; and, finally, the ordeal of the journey and the trauma of adaptation to a new place. His observations suggested not only a direct connection between the modalities of labour recruitment in India and the high incidence of insanity among coolies in British Guiana, but more generally a causal relationship between the experience of migration and the manifestation of mental illness. He also suggested that part of the disparity in the number of natives and immigrants among the inmates was the result of the different social organisation of the two groups in the colony, with the native population often being looked after by relatives or friends, while the immigrants were more likely to be alone and to fall under the care of the asylum. However, Grieve strongly believed that 'the change of circumstances, the separation from country and friends to which the immigrant is necessarily subjected' were key factors in the emergence of mental illness. He implicitly charged the recruiters in India with misconduct, suggesting that 'it is amongst those who possess a tendency to the insane neurosis, those who are mentally unstable that the emigration agent most easily finds his recruits'. Even when signs of insanity were not manifest at the time of recruitment, Grieve believed that those individuals were in such a vulnerable state that the circumstances of migration quickly pushed them 'into the undoubted territory of insanity'. He supported his claims by pointing to 'the number of coolie immigrants who go mad on the voyage hither or within a very short period of their arrival in the colony'.[48]

The fact that 'many either already insane or on the brink of insanity'[49] found their way to the colony through the unscrupulous recruitment conducted in India had already been observed and condemned in 1870 by the Royal Commission of Enquiry investigating the treatment of immigrants in British Guiana.[50] Interviewed by the Commissioners, the protector of immigrants James Crosby had also declared that in several cases, as he inspected the new arrivals, he had had to send immigrants back to India on account of their mental state: 'Last year I sent back two women on one ship. They were in fact idiots, and were sent back. I have done so on more than one occasion'.[51] Even though the Commissioners linked the importation of unfit immigrants to a reduction of suitable labour in the colony and therefore to economic loss, the colonial government took no effective action.

Just as Donald noted how malnourishment was likely to lead to admittance to the asylum, both Grieve and the Commissioners paid great attention to the diet of patients in British Guiana. The diet observed in the Berbice Asylum reflected the dietary scale for hospitals established by a circular of 5 November 1859. However, the evidence collected by the Commissioners showed that the same dietary was not generally adhered to in other institutions in the colony because it was considered too expensive, costing 8d a day for each patient.[52] Grieve circumvented the economic burden on the asylum of patients' diet with occupational therapy; a key aspect of his management was the implementation of a sophisticated system of industrial and agricultural employment, to help patients to develop a sense of responsibility and discipline. By 1881 the asylum operated a farm, a bakery, a printing office, a laundry, a sewing room and a work room for the production of chocolate and spices on its premises. The institution also employed a carpenter, painter, tailor and shoemaker as attendants; each worker was to follow their trade with the assistance of patients. The work carried out by the inmates earned a considerable monetary profit for the asylum, which led to almost complete financial independence for the institution. Already in 1876, the expenses of the institution were covered almost entirely by the Industrial Fund, 'formed from the proceeds of the labour of the patients with a small amount received from the board of paying patients, and the fines and stoppages from the attendants'.[53] Unconcerned with the cost of provisions, Grieve repeatedly remarked on the importance of a 'full and liberal supply of food in the treatment of insanity'[54] and observed its beneficial effects in the asylum, especially among the 'coolies'; he was also keen to stress the good quality of the beef used in the asylum, 'nothing but steers being killed and care is taken in the selection of the animals'.[55] The weight of each patient was registered upon admission and again on discharge, a practice which contributed to the belief that recovery from insanity was always accompanied by an increase in body weight.[56] At the same time, the diet in the institution consisted of 'a fair proportion of animal food, butcher's meat of some kind given every day', making no allowance for religious or cultural beliefs on food.[57] As Grieve himself noted, 'although such a large proportion of the inmates are natives of India', the asylum diet 'contained a great amount of animal food', but '[t]o the eating of beef the coolie soon becomes accustomed and his appearance shows before long how well it agrees with him'.[58]

Use of narcotics

Approaches to mental health in the asylums of Trinidad and British Guiana were also informed by the misuse of substances, their regulation under legislation and their role in furthering a medical understanding of general mental well-being in the nineteenth century. 'Ganga' or hemp was considered a key contributing factor to asylum intake; fluctuations in the number of East Indian patients in Trinidad was ascribed by George Seccombe, Superintendent of the Belmont Lunatic Asylum, to effective implementation of the Ganga Ordinance of 1885. This measure raised the price of the opiate, leading Seccombe to conclude that:

> Ganga now, to the majority of the Coolies must be considered a lux-ury at $10 per lb. compared with 12 cents, the price of the same amount a few years ago. Ganga, though not cultivated in this Colony, is grown, I am informed, in considerable quantities in Grenada, and frequently smuggled across to this Island. It is to be regretted that measures are not taken to prohibit the sale of Ganga, other than at the Drug Store, where it should be treated in the same manner as any other poisonous drug.[59]

Seccombe's position on the significance of drug use and legislation in affecting mental health and psychiatric practices was symptomatic of broader concerns shared by his colleagues across the Empire.

Similarly, Grieve's writing indicated a link between insanity, violence and the immigrants' intemperate habits, which was being established by several colonial doctors during the second half of the nineteenth century through the redefinition of the effects of common drugs on the human brain. In the wake of contemporary European currents of thought, aimed at asserting medicine's jurisdiction over madness, Grieve firmly believed that moral agents, however important in the emergence of mental illness, were generally subordinate to physical determinants.[60] In his *Asylum Journal* he extensively discussed the connection between the physical debilitation produced by the abuse of intoxicating sub-stances and the high incidence of insanity in British Guiana. In a long article entitled 'Narcotics as Causes of Insanity', Grieve identified four major drugs diffused in the colony, namely hemp, alcohol, opium and tobacco, and classified them according to their alleged incidence and impact in cases of insanity. He maintained that each of them caused marked physiological effects upon the human frame and, more espe-cially, on the brain functions. However, while the first two – hemp and alcohol – were particularly potent in this respect, the others – opium and tobacco – were comparatively innocuous.[61]

Indeed, intemperate use of Indian hemp was indicated by Grieve as one of the leading causes of insanity among East Indians in British Guiana. Of the male East Indians admitted to the Berbice Asylum, 70 per cent were found by Grieve to be addicted to ganje smoking, 'a most prolific source of lunacy' in the colony.[62] In September 1881, he produced a further lengthy piece on the theme of 'Insanity from the Use of Ganje' and invoked the lunatic asylums in India as a source of evidence in support of his argument. Equating cultural/ethnic parameters with chemical/biological ones, he maintained that many cases were to be found in the Indian asylums of insanity resulting from the abuse of various preparations of hemp, 'the favourite intoxicant of the Hindoo'.[63] This fondness for the narcotic, explained Grieve, was preserved by the Indian immigrants in British Guiana, most of whom indulged in it so much that the amount they smoked was 'only limited by their power of purchase'.[64] As a consequence, Grieve observed, 'disease of the brain dependent on Indian hemp' was seen very frequently in the Berbice Asylum and possessed very distinctive characteristics.[65] His notes on the effects of hemp were corroborated by several case studies reported in *The Asylum Journal* and he coined the word 'cannabism' to denote parallels with alcoholism. Stereotypes regarding the typical ganje user, who was generally male and Indian, were created and diffused at this time through similar studies. A well-constructed piece of colonial propaganda, written in 1893, explicitly associated characteristics of indolence, idleness and violence attributed to Indian workers with the use of hemp and with insanity. The authors, physician T. Ireland and S. Edinburgh, Government Medical Officer for British Guiana, wrote:

> The habitual smoker is usually an inveterate liar, and like a drunkard, at first attempts to conceal the habit to a certain extent, though it may eventually become known to all his neighbours. As he gradually becomes more addicted to the vice and indulges in it more openly, the first symptoms of mental derangement begin to show themselves. He becomes idle and careless, he neglects his field work, his earnings decrease and his diet naturally becomes more scanty.[66]

Those who refused to work were a dangerous destabilising element in a context where the wealth of the ruling minority could only be guaranteed by the Indian coolies' performance on the plantation. By classifying them as insane, colonial authorities dismissed their claims and disempowered their protest, whereas excessive use of hemp could well be understood as a direct response to the oppressive conditions of indentureship and the inhuman terms imposed by the planters on their workers. Grieve's efforts to assess the effects of 'cannabism' led him to establish an interesting relationship between the use of hemp and the numbing of pain; he explained that the narcotic caused 'a development of nervous energy which finds vent in rapid and excited motion whilst at the same time there is indifference to pain .... The coolie appreciates this action and when he wants to nerve himself to some act of violence he too often resorts to ganje for a supply of the East Indian equivalent for Dutch courage as well as to drown reflection'. While still validating the characterisations above, he also seemed to contextualise the use of hemp in distinctive ways, relating it to the need for nerve or for evasion, both compatible with the oppressed condition of the indentured population.

Another popular means of escape among East Indian immigrants was alcohol, described by Grieve as the most diffused narcotic in British Guiana. Though early temperance reformers had concluded that alcohol had the potential to destroy mental faculties, unlike hemp, it was well-known to the British public and its excessive use was largely tolerated at home and in the colonies. In the emerging field of psychiatry, the abuse of alcohol, previously considered a *result* of insanity, came to be almost unanimously condemned as a leading *cause*. Grieve's position on the subject mirrored this shift in the perception of alcohol from symptom to cause. Analysing the data derived from the medical certificates drawn up on the admission of patients to the Berbice Asylum, he observed that of the men sent to the asylum between 1876 and 1881 nearly one half had been 'previously intemperate to such a degree as to be known as drunkards'.[67] This led him to conclude that 'whatever the nature of the connection between drink and insanity may be, their union cannot be doubted and addiction to its intemperate use now and for generations past goes a great way in explaining the present large amount of madness'.[68] Grieve saw confirmation of this hypothesis in the fact that excessive drinking was prevalent among East Indian immigrants, who made up the majority of the inmates of the asylum. Unlike 'cannabism', alcoholism was considered to be not only responsible for impairing the drinkers' intellectual faculties, but also for having a strong impact on the psychological development of their descendants. 'Alcoholic intemperance', explained Grieve, 'acts as a cause of insanity by predisposing the individual to the disease by exciting it, and still more by producing in the drunkard's descendants the hereditary taint of the insane temperament'.[69] In the broader colonial context, the accent on biological inheritance proved useful in constructing and maintaining the association between intemperance, insanity and race.

Conclusion

The establishment of the asylum system in the post-emancipation British Caribbean was part of a wider process concerning the provision of medical care for the free population of the region. As they arrived in the colonies, British-trained doctors were faced with conditions of exploitation and deprivation, plantation management, substance abuse and the exacerbating psychic anguish of migration and this article has explored how they reacted and adapted to those issues. Seccombe, for example, campaigned for years for the improvement of mental health facilities and provisions in Trinidad, ultimately succeeding in persuading the local government to invest in a new asylum which opened at St Ann in 1900. In British Guiana, Grieve extended his authority beyond the asylum and contributed enormously to the sanitary improvement of the colony. In 1886 he was promoted to the post of Surgeon General and moved to Georgetown, the capital, where he became responsible for the welfare of the entire rural population and for the administration of estate hospitals in the colony. During the later stages of his career he became involved with wider issues of colonial government, public health and immigration, his professional opinion often conflicting with local political and economic interests.

At the same time, the doctors were in a unique position to import knowledge to the colonies, and also to create knowledge on the basis of their experience with colonial patients, contributing to the construction of concepts of alterity and identity, linking patients' susceptibility to mental illness to biological, as well as moral, factors. The high rates of insanity among immigrants in British Guiana were a reality attributed to a combination of different causes, including the trauma of exile, overwork, malnutrition, violence and attempts to acclimatise in new environments far from home. British colonial propaganda often saw it as convenient to blame the abuse of intoxicating substances, particularly ganje and alcohol, for the psychological disorientation experienced by many indentured workers and, at the same time, exploited the diagnosis of mental illness and substance addiction to dismiss workers' dissent and resistance to the indentureship system.

To various degrees, Grieve, Donald and Seccombe detected a relationship between economic exploitation, social vulnerability, substance abuse, malnourishment, the condition of migrancy (no friends, no family, uncertain future, new food/climate, etc.) and the emergence of mental disorders. In some cases, their observations directly linked insanity to the condition of the migrant indentured worker, uprooted from a familiar environment and transplanted into a setting of hard labour, unfair legislation, gender imbalance, and social and cultural isolation. At the same time, they used medical discourses of intemperance and heredity which served to underpin – intentionally or not – colonial views of racial savagery and unruliness, further entrenching the imperial production of race through pathologising it.

Notes

1. Apprenticeship was set at four years for household slaves and six years for field slaves. For further details, see James Latimer, 'The Apprenticeship System in the British West Indies', *The Journal of Negro Education*, 33:1 (1964), 52–7.
2. Over 84 years, from 1834 to 1918, the total inflow into the Caribbean amounted to 536,000 people, 2 per cent of the total entering the United States from Europe between 1831 to 1913. See G.W. Roberts and J. Byrne, 'Summary Statistics on Indenture and Associated Migration Affecting the West Indies, 1834–1918', *Population Studies*, 20 (1966), 125–34.
3. Juanita De Barros, 'Dispensers, *Obeah* and Quackery: Medical Rivalries in Post-Slavery British Guiana', *Social History of Medicine*, 20 (2007), 243–61, 244.
4. K.O. Laurence, 'The Development of Medical Services in British Guiana and Trinidad 1841–1873', in Hilary Beckled and Verene Shepherd (eds), *Caribbean Freedom* (Kingston: Ian Randle, 1993), 269–73, 269.
5. British Parliamentary Papers, Report of the Commissioners Appointed to Enquire into the Treatment of Immigrants in British Guiana, 1871 [C.393] [C.393-I] [C.393-II] XX.
6. Laurence, 'The Development of Medical Services', 272.
7. British Parliamentary Papers, The reports made for the year 1847 to the Secretary of State having the Department of the Colonies: in continuation of the reports annually made by the governors of the British colonies, with a view to exhibit generally the past and present state of Her Majesty's colonial possessions, and of the United States of the Ionian Islands. Transmitted with the blue books for the year 1847, 1847–48 [1005] XLVI, 176.
8. Roberts and Byrne, 'Summary Statistics', 125.
9. Laurence Brown, 'Experiments in Indenture: Barbados and the Segmentation of Migrant Labor in the Caribbean 1863–1865', *NWIG: New West Indian Guide/Nieuwe West-Indische Gids*, 79:1–2 (2005), 31–54, 31–2.
10. Claude Levy, *Emancipation, Sugar, and Federalism: Barbados and the West Indies, 1833–1876* (Gainesville: University Presses of Florida, 1980).
11. The population density of Barbados in 1871 was 906 persons per square mile, compared to 284 in Antigua, 93 in Trinidad and 25 in British Guiana. The number of agricultural workers greatly exceeded demand and the price of land was prohibitive; 'in British Guiana and Trinidad it was being occupied freely by squatters', in Barbados prices were as high as £500 per acre. See Levy, *Emancipation, Sugar, and Federalism*, 79, 101, 135.
12. Levy, *Emancipation, Sugar, and Federalism*, 80.

13. Brown, 'Experiments in Indenture', 48, 50. During the 1860s new milling technology concentrated the harvest in British Guiana and Trinidad, which generated new opportunities for seasonal migration. However, with the economic depression that hit the sugar industry in 1884, planters were forced to conclude that Barbadian immigration was no longer profitable and controlled immigration plans were abandoned.

14. British Parliamentary Papers, The reports made for the year 1847.

15. Anne-Marie Lee-Loy, 'Introduction', to Sir Cecil Clementi, *The Chinese in British Guiana* (1915, republished Coventry: Caribbean Press, 2010), ix–xxiv, xii.

16. Clementi, *The Chinese in British Guiana*, 22.

17. This movement of Chinese labourers into the West Indies was part of a larger nineteenth-century worldwide migration from southern China to South Africa, North and South America, and Australia. See Lee-Loy, 'Introduction', xiv.

18. British Parliamentary Papers, British Guiana and Mauritius. Copies of all orders in council, or colonial ordinances for the better regulation and enforcement of the relative duties of masters and employers, and articled servants, tradesmen and labourers, in the colonies of British Guiana and Mauritius, and of correspondence relating thereto, 1837–38 (180) (232), LII.

19. John Scoble, *Hill Coolies: A Brief Exposure of the Deplorable Conditions of the Hill Coolies, in British Guiana and Mauritius, and of the Nefarious Means by which they were Induced to Resort to these Colonies* (London: Harvey and Darnton, 1840), 9.

20. Peter Ruhomon, *Centenary History of the East Indians in British Guiana, 1838–1938* (1947, republished Georgetown: East Indians 150th Anniversary Committee 1988), 127.

21. Scoble, *Hill Coolies*.

22. British Parliamentary Papers, Report of the Commissioners, 1871, 59.

23. David Dabydeen, 'Introduction', to Edward Jenkins, *Lutchmee and Lutchmee & Dilloo: A Study of West Indian Life* (Oxford: Macmillan Caribbean, 2003), 4.

24. The inhuman conditions which marked the transportation of Indian immigrants to the West Indian colonies caused high mortality during the journey. Of the 2,652 emigrants to British Guiana 87 or 3.28 per cent died; among the 1,433 emigrants to Trinidad 46 or 3.21 per cent died. The total number of deaths among the 4,085 emigrants conveyed to the two colonies was 133, or 3.25 per cent. The same report provides numbers for emigration from the United Kingdom to the United States and British North America. Of these 85,555 travelled in steamers and 74,208 in sailing vessels. The mortality for the former amounted to 0.06 per cent, the latter to 0.44 per cent: British Parliamentary Papers, Emigration Commission. Twenty-fifth General Report of the Emigration Commissioners, 1865 [3526], XVIII.

25. Ron Ramdin, *The Other Middle Passage: Journal of a Voyage from Calcutta to Trinidad, 1858* (London: Hansib, 1994), 11.

26. Ibid., 30.

27. Marina Carter and Khal Torabully, *Coolitude: An Anthology of the Indian Labor Diaspora* (London: Anthem Press, 2002), 164.

28. British Parliamentary Papers, Report of the Commissioners, 1871, 3.

29. Ved Prakash Vatuk, 'Protest Songs of East Indians in British Guiana', *The Journal of American Folklore*, 77:305 (1964), 220–35, 227.

30. Kusha Haraksingh, 'Control and Resistance among Indian Workers: A Study of Labour on the Sugar Plantations of Trinidad 1875–1917', in David Dabydeen and Brinsley Samaroo (eds), *India in the Caribbean* (London: Hansib, 1987), 61–77, 63.

31. The construction of the asylum was delayed by the colony's economic recession. The returns of prisoners for the year ending 31st December 1858 shows that 5 people were committed for insanity. See British Parliamentary Papers, The reports made for the year 1858 to the Secretary of State having the Department of the Colonies; in continuation of the reports annually made by the governors of the British colonies, with a view to exhibit generally the past and present state of Her Majesty's colonial possessions. Transmitted with the blue books for the year 1858. Part I – West Indies and Mauritius, 1860 [2711] XLIV.

32. British Parliamentary Papers, The reports made for the year 1860 to the Secretary of State having the Department of the Colonies; in continuation of the reports annually made by the governors of the British colonies, with a view to exhibit generally the past and present state of Her Majesty's colonial possessions. Transmitted with the blue books for the year 1860. Part I. – West Indies, Mauritius, and Ceylon, 1862 [2955] XXXVI and *Report on the Belmont Lunatic Asylum* (Trinidad, 1899).

33. *Annual Report of the Surgeon General on the Medical Service and Medical Institution of the Colony for the Year 1889* (Trinidad, 1890).

34. A surprisingly low number of Barbadians entered the asylum in 1887, compared to the average of 20 Barbadian admissions in adjacent years. About one-fourth of inmates were natives of the colony, followed by comparatively small numbers from Africa, Barbados, China, Dominica, Grenada, Ireland, Montserrat, Nevis, St Vincent, Tobago, Venezuela and the United States. *Annual Report of the Surgeon General on the Medical Service and Medical Institution of the Colony for the Year 1887* (Trinidad, 1888).

35. Henry Light was Governor of British Guiana between 27 June 1838 and 19 May 1848.

36. Sixty of those patients were readmitted several times, 'One having been admitted eight times; one, four times; nine, three times; and 32 twice'. This reduces the actual number of patients admitted to 135. British Parliamentary Papers, The reports made for the year 1846 to the Secretary of State having the Department of the Colonies: in continuation of the reports annually made by the governors of the British colonies, with a view to exhibit generally the past and present state of Her Majesty's colonial possessions, and of the United States of the Ionian Islands. Transmitted with the blue books for the year 1846, 1847 [869] XXXVII.

37. James S. Donald, 'Notes on Lunacy in British Guiana', *Journal of Mental Science*, 22 (1876), 76–81, 76.

38. Ibid., 76–7.

39. Ibid., 77.

40. Ibid., 78.

41. Ibid., 79.

42. Ibid., 78.

43. British Parliamentary Papers, Papers relating to Her Majesty's colonial possessions. Reports for 1879, 1880 and 1881. (In continuation of [C.–3094.] August 1881), 1882 [C.3218] XLIV.

44. James Mills, *Madness, Cannabis and Colonialism: The 'Native Only' Lunatic Asylums of British India 1857–1900* (Houndmills: Palgrave Macmillan, 2000), 16–21.
45. Robert Grieve, *The Asylum Journal for 1881–1886* (Berbice: The Asylum Press, 1882–86). Reprinted in two volumes with an introduction by Letizia Gramaglia (Coventry: The Caribbean Press, 2010); all references in this paper are to the 2010 edition.
46. Grieve, 'Five Years Insanity', *The Asylum Journal* (No. 3, 1881), 34.
47. *British Guiana Ordinance No. 4, 1864. An Ordinance to Consolidate and Amend the Law Relating to Immigrants, Part XIII of the Medical and Sanitary Care of Immigrants.*
48. Grieve, 'Five Years Insanity', 35.
49. Grieve, *The Asylum Journal* (No. 6, 1881), 72.
50. British Parliamentary Papers, Report of the Commissioners, 1871, 3.
51. Edward Jenkins, *The Coolie His Rights and Wrongs: Notes of a Journey to British Guiana, with a Review of the System and of the Recent Commission of Inquiry* (London: Strahan, 1871), 227. See Alison Bashford's article in this volume for medical inspections and the rejection of lunatics at borders.
52. British Parliamentary Papers, Report of the Commissioners, 1871, 152–3.
53. Grieve, *The Asylum Journal* (No. 1, 1881), 15.
54. Ibid., 14.
55. Grieve, *The Asylum Journal* (No. 10, 1881), 127.
56. Grieve, 'Food and Insanity', *The Asylum Journal* (No. 25, 1883), 321–2.
57. Grieve, *The Asylum Journal* (No. 10, 1881), 119.
58. Grieve, 'Food and Insanity', 323.
59. *Annual Report of the Surgeon General on the Medical Service and Medical Institution of the Colony for the Year 1889* (Trinidad, 1890).
60. Mills has asserted that 'Throughout the nineteenth century medical men in Europe were struggling to assert their authority over the psyche. In other words doctors needed to prove that the brain and its workings were properly their concern and not the concern of other professional groups like the clergy who could claim specialist knowledge of the routes to psychological well-being': Mills, *Madness, Cannabis and Colonialism*, 51.
61. Grieve, 'Narcotics as Causes of Insanity', *The Asylum Journal* (No. 11, 1882), 132–5.
62. Ibid., 134.
63. Grieve, 'Insanity from the Use of Ganje', *The Asylum Journal* (No. 7, 1881), 82.
64. Ibid.
65. Ibid.
66. Thomas L. Ireland and S. Edinburgh, 'Insanity from the Abuse of Indian Hemp', *The Alienist and Neurologist*, 14 (1893), 622–30, 622.
67. Grieve, 'Intemperance and Insanity', *The Asylum Journal* (No. 6, 1881), 73.
68. Ibid.
69. Grieve, 'Narcotics as Causes of Insanity', 132–3.

4
The Colonial Travels and Travails of Smallpox Vaccine, c.1820–1840

Katherine Foxhall

Introduction

In 1803, the newly established Royal Jennerian Society expressed confidence that the complete extermination of smallpox was under their control. 'It is not in the course of human probability that centuries will again present such an opportunity of doing good', it declared.[1] Five years earlier, Edward Jenner's *An Inquiry into the Causes and Effects of the Variolae Vaccine* had demonstrated that deliberate exposure to the infection of cowpox conferred permanent protection from the much more serious and contagious disease of smallpox. This process, developed in the fields of rural England, seemed safer and decidedly more modern than the practice of inoculating with live smallpox matter that had been introduced to Britain earlier in the eighteenth century.[2]

Jenner's vaccine moved quickly. Missionaries, military surgeons, merchants, colonial officials and elite travellers carried the precious packages in their luggage. Cowpox matter soon arrived in the Mediterranean, Russia, North America and Brazil. Early publications gave detailed instructions on its use. The vaccinator would puncture the cowpox vesicle with a lancet and transfer the matter directly from one recently vaccinated person's arm to the next recipient. This was known as the arm-to-arm method, by which vaccine arrived in Bombay in 1802, utilising a relay of children across land from Baghdad.[3] Alternatively, matter could be preserved for travel on the tips of quills, ivory tips or toothpicks. Preferably, matter should be dried and protected between two sheets of glass wrapped in paper and revived by moistening with a little water when needed.[4] In 1803, Governor Gidley King requested vaccine matter for New South Wales; he was rewarded the following year when the *Coromandel* delivered a supply from the Royal Jennerian Society. The

British were not alone in these endeavours; the Portuguese and French also transported vaccine across the Indian, Pacific and Atlantic Oceans.[5] In 1803, the Spanish Crown sponsored a Royal Maritime Vaccination Expedition which took Jenner's vaccine to Puerto Rico, Guatemala and then on to the Philippines.[6]

Jenner's vaccine, Mark Harrison has argued, 'was undoubtedly the most tangible example of medical progress during the eighteenth century'.[7] Historians have charted in detail this early transmission of vaccine matter from the fields of rural England.[8] We know a great deal about the men and women whose personal contacts, private interests, Enlightenment ideals of humanitarianism and progress and military concerns introduced cowpox vaccine to far flung parts of the world.[9] However, this chapter argues that we have taken too much for granted in these early expressions of confidence in vaccine's 'global' spread because we still know very little about how vaccination practices and matter were made to work in particular places and situations in the first decades of the nineteenth century.[10] Historians have lost vaccination's thread around 1810 and assumed the inevitability, rather than researched the circumstances, of its spread from then onwards. We have not questioned why, on returning to the subject in the second half of the nineteenth century, we have found a distinctly less triumphant story. As governments began to make inoculation illegal and vaccination compulsory through national and imperial acts, Nadja Durbach has shown that anti-vaccination sentiment became a central theme to Victorian 'body politics' in Britain, while for India David Arnold has demonstrated that opposition to government-driven vaccination programmes revealed a deep distrust of British rule, showing 'how readily state medical intervention was identified with other coercive and alien aspects of the colonial regime'.[11]

There is a profound disjuncture between these two vaccine stories in the nineteenth century; how does a history of global migration intersect with local, historically specific 'body politic' scales of vaccine's history? Two observations about vaccine have proved particularly useful in considering this question. First, Melissa Leach and James Fairhead have observed that although 'vaccination is easily represented as a universal, neutral good, it is actually deeply bound up with politics: with struggles over status, authority and value'.[12] Second, Alison Bashford has suggested that we might see vaccine as 'a kind of colonial contagion', a deliberately introduced bodily contaminant 'integrally related to local and global migrations and to a history of travel, orientalism and colonialism'.[13] Here, I want to push these colonial and bodily observations further, to reconnect with specific cases of vaccine's travels in the first half of the nineteenth century. What were people doing with and saying about vaccine? What happened in the middle ground between local and global? What work had to be done by and on behalf of vaccine matter, to ensure its success? What groups of people in different places were utilised in this process that made health, as well as illness, migrate?

This chapter considers two specific cases in the British context: the vaccination by naval surgeons of Irish child migrants who sailed to Quebec in 1825, against a background of vaccination on convict vessels en route to Australia; and the Governor of Barbados' correspondence regarding vaccine supplies in the years following slave emancipation. Together, these very different examples demonstrate that we simply cannot understand vaccine's global spread as a smooth process that occurred through philanthropic motives. Rather, we should recognise that vaccine succeeded because it became a colonial and political commodity which could serve contemporary colonial imperatives and changing patterns of migration, both free and forced, in the early nineteenth century. If the themes of health and migration are explicit in this chapter, ethnicity plays a more subtle role, but it appears in two ways. In the first case, we consider a group of Irish migrants, mainly children, whose ethnicity is not coincidental to their being used as experimental vaccine carriers. Secondly, we note that though Jenner's vaccine was explicitly coded as 'English', in contrast to earlier practices of inoculation that had been introduced from the Ottoman Empire, the late nineteenth-century concerns about the 'purity' (i.e. English origins) of vaccine matter seem not to translate easily back to the first half of the nineteenth century. In this earlier period, I suggest, vaccine was so unreliable that its value came simply from being 'live', and thus was equally as acceptable whether it came in 'pure' form or had passed through Irish, convict or African bodies. Only once vaccine's value had been established beyond doubt would the question of purity, or of 'clean lineage', come to the fore later in the century.[14] While we have long known that vaccine relied on 'little boys' for its odyssey, the two examples this chapter examines suggest that we need to think more expansively about how surgeons used marginal and disenfranchised migrants in order to establish vaccine's viability, and how, in doing so, vaccine helped inform ideas about the political and practical functioning of empire in a key period in imperial history.[15]

Vaccine uncertainties

At the turn of the nineteenth century, physicians fell over themselves to evangelise Edward Jenner's discovery.[16] From the beginning, military surgeons had proved some of the most enthusiastic subscribers to Jenner's gospel. In his famous *Medicina Nautica* (first published in 1797) the naval surgeon Thomas Trotter advocated a 'general inoculation' of sailors in ships and fleets.[17] Two years later, on returning from sea, Trotter found medical attention fixated by Edward Jenner's discovery; such evidence, Trotter believed, was 'pregnant with wonders' and he declared his wish to see some of Jenner's Gloucestershire cows transferred to the navy farm for the inoculation of seamen.[18] The Admiralty's Sick and Hurt Board undertook its own trials before ordering that vaccination be made available to sailors, and in 1801 awarded Jenner a gold medal as a token of their appreciation for his discovery.[19] In Ireland, too, early reports from Dublin's Cow Pock Institution (established in 1804) showed that demand from Army surgeons was consistently high, while regular practitioners were much slower to adopt the practice.[20]

Although the Navy adopted vaccination to prevent smallpox outbreaks at sea, the practice remained voluntary for sailors throughout the first half of the century. However, from 1815 naval surgeons were given responsibility for the medical care and discipline of convicts bound for the Australian colonies. At the end of each voyage – in order to secure their pay – surgeons were required to submit an official journal of the medical events of the voyage, many of which record that surgeons performed vaccinations on prisoners in their care. At first, the time and space of long voyages had allowed interested surgeons to experiment idiosyncratically. By the 1820s, the Commissioners of the Navy were routinely sending packages of vaccine virus from the National Vaccine Establishment in London with surgeons on ships to New South Wales. Their instructions required surgeons to vaccinate convicts who could not demonstrate that they had either had smallpox or been vaccinated.[21] Through the vaccination records of thousands of men, women and children, reports of vaccination in these journals provide clear evidence of the persistent difficulties of preserving and transporting vaccine.

Irish emigrants

Nine years after naval surgeons took responsibility for convict voyages, the British government employed them to superintend another migrant group. In 1825, nine ships sailed from Cork in Ireland to Quebec, Upper Canada, with 2,024 Irish Catholic labourers and their families. Most of the emigrants, selected from over 50,000 applicants, came from parishes in the Blackwater River Valley in County Cork, and were sponsored by local landlords. The majority travelled as families with an average of four or five children.[22] Funded by the Colonial Office, these journeys were one of the earliest experiments with state-sponsored emigration and colonisation from the British Isles. In their journals – the same kind as they submitted for convict voyages – the naval surgeons talked of these voyages as a colonial experiment and were conscious that their duties went beyond medical care. They observed how the emigrants coped with working in extreme heat; one surgeon recorded that during the journey up the St Lawrence River from Lachine they had worked in the mornings and evenings to avoid the midday sun. Some of the children had attacks of cholera, while women became feverish for a day or two from exposure to the sudden changes in weather and the high temperatures.[23] The surgeons also offered the government advice about altering food rations and supplying extra medical comforts should any future attempts at such co-ordinated migrations occur.

The vaccination of the emigrants and their children was a crucial aspect of this enterprise. It offered the means to protect the health of a colonising group and to deliver fresh vaccine to Quebec. Six of the eight surgeons whose journals survive made specific comment. Decisions about how to use the vaccine supplied from London had apparently been left to their own discretion. On the ship *Fortitude,* surgeon Connin did not attempt to get live vaccine to Quebec. He performed all the 29 vaccinations of children and adults in the first few days of the voyage, 'the whole of [which] proved unsuccessful'. The surgeon of the *Elizabeth* similarly vaccinated 14 children in Cork Harbour while waiting for a favourable wind to sail on 18 May.[24] Again, all failed. On the *Amity,* surgeon McTernan reported (after signing his name, and then remembering to add his thoughts on vaccine) that he also made repeated trials of the vaccine matter, but without effect and 'as far as I can learn it failed throughout'. McTernan commented that the same failure had occurred before while he served on a convict ship.[25]

On the *Albion,* John Thomson tried a different approach. He vaccinated three children on 6 May 1825 while they waited in the harbour for favourable weather to begin the Atlantic crossing. Eight days after the procedure – the generally accepted length of time a vaccine took to produce the distinctive full 'areola' in the patient's arm to indicate success – the vaccine had achieved some results, but had failed in all three children. Thomson chose one of the children, a seven-year-old girl named Catherine Barry whose pock had failed only on the last day, and tried again. Into the skin on her right arm he again inserted the vaccine sent to him by the Navy Board. In the left arm, however, Thomson vaccinated Catherine with an alternative supply that had been sent to him by Dr Johnstone, the secretary of the combined London Vaccine Institution and Royal Jennerian Society in London, first founded in 1803 and funded entirely by private subscription.[26] Again, the 'official' Navy vaccine failed, but Dr Johnstone's matter produced a 'very fine' pock in Catherine's left arm. With renewed confidence, Thomson now vaccinated Catherine's two younger siblings with the alternative vaccine, as well as three other children. Nine days later (in mid-Atlantic) the surgeon chose the boy whose arm had produced the best results – Catherine's five-year-old brother John – and used him to vaccinate another eight children. He repeated the process twice more, each time choosing one of the most recently vaccinated children as the source of his fresh supply. He undertook the final procedure on 11 June, just four days before the *Albion* arrived in Quebec. While surgeons on convict ships in this period often commented that they encountered resistance, particularly to the vaccination of children, the surgeons of these vessels to Canada made no comment about how the children, or their families, understood or reacted to these events. It is highly likely these Irish children from labouring families rarely spoke English, and thus received little in the way of explanation. The surgeons were more interested in the vaccine, and one surgeon at least simply considered the children to have been 'living subjects'. 'I have to remark', Thomson concluded, 'that the vaccine virus supplied me was effete, but Dr Johnstone of Burr Street having sent me some, I was enabled to take it out in the living subject to Quebec'. Thomson emphasised the total failure of the Navy Board's supplies in his and other ships, pointing out that this situation 'might have been attended with unpleasant consequences as many of the children had only just recovered from the small pox & it was doubtful if some of the families might not have brought the disease with them'.[27] We will return to the significance of this failure shortly, but Thomson's description of the emigrants as 'living subjects' emphasises their apparent passivity; their very susceptibility to disease transformed children into vessels for transporting vaccine supplies to Quebec.

The surgeons accompanying the Irish settlers who went to Upper Quebec in 1823 and 1825 certainly considered that they were engaged in a colonial experiment, but there is another relevant precedent. In 1810, members of the Royal College of Surgeons in Ireland reported with great enthusiasm about an 'experiment' on 19 of the Dublin Foundling Hospital's children which had proved – by injecting them with live smallpox matter – that vaccination's power did not reduce over time. A. Bailie of the Cow Pock Institution clearly felt that this was of import beyond Dublin. He ordered 3,000 copies of the report to be printed and distributed throughout the United Kingdom; the experiment was also widely publicised in the *Medical and Physical Journal* and the *Philosophical Magazine*.[28] As other contributors to this volume mention, historians have long argued about Ireland's colonial status in the early nineteenth century. As Stephen Howe has commented, the idea that Ireland was a 'testing ground' for Britain's policies is more often the basis for 'sweeping affirmation and denial' than sustained investigation.[29] While we should not therefore, on the basis of these examples, draw a hasty conclusion that Irish children were exceptional, it is evident that experimental vaccination procedures with Irish children had previously been used to provide evidence in debates about the practice's long-term viability and efficacy. If in 1809 the questions had been about vaccination's permanent protection of individuals, by 1825 findings on Irish children informed the British government's early attempts at running schemes of mass assisted emigration.

This episode also suggests that we need to think more expansively about the role of migrant children in colonialism. Although by the 1830s, naval surgeons would complain frequently about high numbers of children on Australian emigrant ships, in part because they were so prone to introducing eruptive and highly infectious childhood diseases such as whooping cough, marasmus and measles, it was precisely this susceptibility that made them the best vaccine carriers; the children of emigrants, convicts and members of the military guard were indispensable to spreading smallpox vaccine around the world. As vaccine incubators, the raison d'être of child migrants was not to be rescued or reformed, to relieve a perceived social problem or even to be 'units of labour'; rather, they acted, quite literally, as medical vessels in the service of colonial settlement.[30] Forming human chains during voyages, children were live extensions to the surgeon's travelling medical supplies.

Failure and rivalry

Unfortunately for many surgeons even the healthiest of unvaccinated children were of no use if their vaccine matter was inert. On arrival in Quebec in 1825, the naval surgeons accompanying this group of Irish settlers discussed their frustrations. No matter when or how the surgeons used their supply, the Navy's vaccine failed in every case. John Tarn was most direct in reporting back on this point: 'the Vaccine Virus supplied by the Navy Board...proved to be wholly inert'. Thirty of the emigrants on his ship had needed vaccination, and although he 'tried the matter in various ways, and in the most cautious manner...in no instance did it take effect'.[31] William Burnie suggested that the Navy's vaccine 'must either have been originally bad or it must have been kept too long'. It had been improperly packaged in a single fold of white paper, and in an 'exceedingly minute' quantity, which could scarcely be seen on the glass.[32]

John Thomson admitted that many of his emigrants had been 'weakly', particularly the children, when they embarked, having previously suffered from sickness and want of nourishment. A key aspect of contemporary debates about vaccination was whether the procedure would work for sickly people. Indeed, when the *Albion* was employed to carry convicts from London to Australia three years later, its surgeon would lament that his attempts at vaccination failed because the convicts were environmentally and constitutionally 'in the most unfavourable circumstances to go thro' the disease'.[33] It is significant, therefore, that in 1825 the surgeons who sailed to Quebec unanimously asserted that it was the Navy's packages – and not the Irish emigrants – that were to blame. Thomson believed that the only vaccine to take effect (Dr Johnstone's) had been successful because it was 'carefully done up in the tinfoil'.[34]

The Navy had been taking its regular supplies from the rival to Dr Johnstone's institution: the National Vaccine Establishment. Founded in 1808, the Establishment was funded by Treasury grants, and its Board's mandate was to organise free vaccinations, distribute lymph and investigate reports of failures. The supporters of these different institutions could be openly competitive, and at times positively hostile towards each other, particularly over matters of funding. Initially, the National Vaccine Establishment had been very successful – its vaccinators reported over 87,000 vaccinations in 1817; by the 1820s, however, its intellectual leadership had declined and there were concerns about instances of failure.[35] Both institutions increasingly derived authority and credibility from supplying international and colonial networks and institutions and lists of foreign recipients occupied a prominent place in the annual reports. Only a few weeks before the departure of the Canadian emigrant ships in 1825, the National Vaccine Establishment had listed Madras and Bengal, Jamaica, Barbados,

New South Wales, Sierra Leone and Buenos Ayres among its recipients and confidently declared that the frequent applications received from abroad were made because their lymph was 'more genuine and efficacious'.[36] Dr Johnstone's 'gift' of a package of vaccine to the naval surgeons embarking on a new colonial experiment in 1825 was thus a timely political act, just as the rival Establishment's confident declarations proved to be empty of substance. By 1830, the annual accounts of the Royal Jennerian Society suggest that the Navy had shifted allegiance somewhat; it now contributed £1,000 for supplies to the Navy and military settlements abroad.[37]

The failure of the National Vaccine Establishment's lymph in 1825 must also be viewed against a background of increasing public doubts about the efficacy of the vaccine. For example, in 1822, though the Edinburgh physician John Thomson remained convinced of vaccine's 'wonderful power', he was concerned that the procedure was 'not in all circumstances an absolute, or even a general preventive of small-pox'.[38] Another writer admitted that while the procedure had initially promised to provide 'a perfect and uniform protection' against smallpox, it now appeared imperfect or partial, and in a few cases seemed to exercise 'little or no influence in arresting or modifying the virulence of the subsequent disease'.[39] By the 1820s and 1830s medical debates about smallpox vaccine centred on questions of method, safety, the permanence of protection, preserving matter and whether cowpox's power deteriorated as it passed through thousands of human carriers.[40] In 1829, Dr Delagrange wrote to the *Lancet* from Paris to ask whether, in England, 'you have remarked the diminution of the anti-variolous property of the vaccine', because in France, 'we observe it every day'.[41] It was becoming increasingly clear that maintaining constant supplies of vaccine matter in all but the most densely inhabited areas was nearly impossible. In 1831, the Board of the government-funded National Vaccine Establishment explained that although they constantly admonished the recipients of lymph to maintain their own supplies, the 'incessant' applications they received suggested that it was impracticable to keep up supplies anywhere but in London, where vaccinators could assist each other.[42]

Throughout the 1820s and 1830s, particularly as they continued to accompany convicts and government emigrants to the Australian colonies, naval surgeons tried again and again to make good lymph from bad, to extract life from inert matter, success from failure. Thomas Logan's expressions of despair in 1828 at his failed attempts to transport vaccine to New South Wales are particularly telling. Logan waited a full four months after leaving England before vaccinating the convicts of the *Albion*. Confessing that he had 'reserved the process...for a late period of the voyage' because he 'wished to land a great supply of matter', he acknowledged that he had attempted 'in short, to do too much' and had 'defeated my own object'.[43]

When surgeons vaccinated at sea, they juggled the desire to arrive with live vaccine lymph with the need to prevent outbreaks of disease during the voyage. By the late 1830s, as government-assisted emigration to Australia was becoming an established practice, these two different priorities would become aligned with two different streams of migrants. Thus, in 1838, surgeons in charge of government-assisted emigrants were instructed to examine the people's arms or see certificates of vaccination and in doubtful cases renew the vaccination before departure, to prevent smallpox breaking out during voyages, a particular risk with so many children on board.[44] Emigrants also received clear warnings that no family would be allowed to embark unless they could provide a certificate from a respectable medical practitioner, that each of their children had either had the smallpox or been vaccinated.[45] By contrast, on convict ships, surgeons were instructed 'to keep up such a succession of vaccinated cases as may enable him to convey fresh virus to the colony, if the number of Convicts or Passengers on board, who may not have had the Small-Pox nor undergone Vaccination, and who shall consent to be vaccinated, will admit of it'. The element of consent was important here, and there is evidence that many convicts refused to be vaccinated, a tactic that was unavailable to emigrants.[46]

Colonial newspapers show just how important deliveries of live vaccine matter were in the colonies. In 1839, the *Sydney Morning Herald* explained that despite repeated attempts, no ships had arrived in the colony with lymph for two years.[47] Thus, when *HMS Pelorus* arrived with live supplies, a deputation of local Sydney 'subscribers' presented the surgeon of the ship, Dr Reilly, with a silver snuffbox, engraved with the words 'a token of regard for the benefit he has conferred by successfully introducing the vaccine lymph into N.S. Wales'.[48] Taking account not just of where the vaccine went, but how it was made to work at a distance suggests that we have greatly underestimated the extent and significance of vaccine's frequent failure during the first half of the nineteenth century. If so, what does it mean that these successes were perhaps the exception, rather than norm? This question also invites us to re-assess our understanding of what was at stake in making vaccine matter into a precious material.

Barbados: Vaccine as political commodity

Colonial vaccine networks mapped onto long-established circuits of trade, people and supplies linking maritime ports and islands. Importantly, vaccine packages did not change hands for financial profit; vaccination was widely talked of as a blessing to be bestowed, not an opportunity for profit.[49] In 1831, a series of exchanges in the *Lancet* made precisely this point. One contributor asserted that if the London Vaccine Institution and Royal Jennerian Society continued to supply vaccine only to their subscribers, rather than to anyone who needed it, they could not 'be considered as "exerting themselves and aiding others to the utmost in the good cause of dispensing the blessings of vaccination"'.[50] Precisely because of its fragility, live vaccine thus became a highly prized political and diplomatic object, the exchange of which conferred status on the giver, and a debt of gratitude on the part of the receiver.[51]

The status of vaccine as a political commodity embedded in the workings of early nineteenth-century colonialism is perhaps clearest in some of the dispatches circulating in the British Caribbean colonies in the years around slave emancipation.[52] On 22 March 1837, Sir Evan John Murray Macgregor, Governor of Barbados and the Windward Islands, wrote to Lord Glenelg, the Colonial Secretary in London, to renew a request for supplies of vaccine lymph that he had made two months earlier. Macgregor assured Glenelg that every precaution had been taken to prevent the introduction of smallpox to the Windward Islands, but the 'formidable disease' was now present in ten nearby colonies, including Grenada, Tobago and Trinidad to the south, and Martinique and Dominica immediately to the north. With his letter, MacGregor enclosed several dispatches he had received in the preceding month.[53] On 7 March, Lieutenant Governor Darling of Tobago had gratefully acknowledged receipt of 'a quantity of Lymph, obtained from subjects recently vaccinated at Barbados', which, Darling insisted, 'could not have afforded a more opportune and agreeable instance of Your Excellency's kind attention to our welfare'. Other letters of thanks came from the Lieutenant Governor of St Vincent, Barbados' nearest island neighbour, and Lieutenant Governor Doyle of Grenada.[54]

First introduced to the West Indies in 1518–19, smallpox had long been a problem related to the traffic of slaves to the West Indies and the maintenance of their health on the plantations.[55] One historian has argued that experimental inoculation practices during eighteenth-century slave voyages had 'provided positive proof of the efficacy of inoculation...where the English aristocracy often feared to tread, the slaves who endured the middle passage provided the ironic and dramatic evidence of the success of a radical medical intervention'.[56] Owners of larger West Indian plantations certainly practised inoculation of slaves from the 1770s, before adopting Jenner's cowpox method in the early nineteenth century. Barbados did not set up a Vaccine Establishment as had occurred on the larger nearby islands of Jamaica, Trinidad or Demerara in the second decade of the century.[57] Instead, by 1837 the Governor of Barbados was distributing packages of vaccine with diplomatic dispatches that passed between the islands and colonies of the region. Since 1833, Barbados had been the seat of a general Windward Islands governorship, but MacGregor's use of vaccine went beyond reinforcing the subordinate position of the islands under his jurisdiction, including St Vincent, Grenada and Tobago.

Despite the island's distance from Barbados, William Rogers Isaacs, President of the Assembly of Tortola, had also received a parcel from MacGregor (at an opportune moment, after two cases of smallpox had occurred on the island). After vaccinating the people of the town, Isaacs now intended to 'retain the matter and to proceed as extensively as may be in vaccination of subjects'. Isaacs concluded by emphasising that the inhabitants of the tiny island of Tortola 'feel, as I do myself, the highest sense of gratitude towards your excellency for this additional mark of Your Excellency's consideration for the welfare of this colony'. On 27 February, Sir James Carmichael Smyth, Governor of Guiana, had also written to MacGregor explaining that there was no vaccine matter to be procured in Demerara, and requesting that MacGregor send 'any that can be spared from Barbados', at the first opportunity.[58]

MacGregor was acting as the hub of a regional network in the Eastern Caribbean, using his position to make vaccine a political and diplomatic commodity with which to extract gratitude from and instil a sense of debt in his fellow governors. Importantly, vaccine travelled through a network of political rather than medical men. It was MacGregor, not the members of the Barbados Board of Health, who controlled the onward movement of the lymph that arrived from London at the end of April.[59] This point is not insignificant in these key years in the context of the collapse of slavery and the establishment of a new and as yet uncertain political and social order in the region.[60]

Subsequent events highlighted the political importance of maintaining these supplies. A little more than two weeks after MacGregor received the fresh lymph in early 1837, he received information that a government ship, the *Harpy,* had captured a slave vessel flying the Portuguese flag off the south-west of Martinique; the *Florida* had 280 Africans on board.[61] Under the terms of the 1817 convention signed with Portugal, the suspected slaver had to be taken back across the Atlantic to Sierra Leone for trial before the British Mixed Commission.[62] To begin with, however, the Africans were landed in St George's, Grenada, where Lieutenant Governor Doyle ordered that they be distributed under indenture to the island's plantations, and communication prevented with the town of St George's where smallpox prevailed. The Collector and Comptroller of Customs were appointed Guardians of all the slaves under the age of 21 and were authorised to provide clothing, food, vaccination and medical attendance for the Africans, amounting to a cost of £80 7s 8d.[63]

The following month, naval commanders and the islands' Governors attempted to use the case of the *Florida* (together with the *Negunha* and *Phoenix* slavers captured the previous year) to demonstrate the necessity of establishing a new Mixed Commission for slavery in the Windward Islands. St George's on Grenada, in particular, offered a 'safe and commodious' harbour for the detention of vessels. If Doyle and MacGregor emphasised the desirability of preventing the expense and inconvenience of sending suspected slavers, accompanied by naval vessels, to Sierra Leone for adjudication, such a development would also help ensure future labour supplies for the plantations now that slavery and the temporary apprenticeship system that replaced it were coming to an end. Having a Commission in the Caribbean, Doyle observed, would allow them to locate the captured Africans 'wherever most advantageous' in the islands.[64] Although ultimately their representations failed, it is evident that the capacity to vaccinate several hundred new arrivals at any one time against smallpox was an essential element of this political manoeuvring.

In the years of emancipation, as plantation owners attempted to ensure that labour costs remained low and ex-slaves and labouring people negotiated the changing and increasingly repressive conditions of a deeply uncertain economic and social future, high levels of mobility and often-clandestine migration occurred in, to and from Barbados.[65] From 1838, this demand for cheap labour would increasingly be met by indentured immigrants from Africa, India and China.[66] Two years after the capture of the *Florida*, in August 1839, the President of the Barbados Board of Health, James Butcher, reported 'a case of eruptive disease of a suspicious character' on the Friendship Estate, one of Barbados' sugar plantations. Butcher, and two other physicians who visited the estate, believed that the woman presented all the symptoms of a case of modified smallpox, meaning that she had already been vaccinated.[67] The Board of Health explained to the Governor that the patient had probably received her infection from a man named Pudmore who had died a fortnight earlier, and with whom she had been 'in constant communication on the Sunday'.[68] Butcher explained that Pudmore had arrived on the island by the *Glengary* with an eruptive disease, which had been reported to be contagious at the time, but no action was taken.[69]

Governor MacGregor was concerned by the *Glengary's* apparent violation of quarantine laws, but he was also greatly displeased with the Board's request that he procure a supply of vaccine lymph from the Principal Medical Officer at the military station in response to the apparent arrival of smallpox with Pudmore. Why had they failed to alert him that Barbados had no vaccine matter of their own? He reminded the Board of his efforts in 1837 to bring the scarcity of lymph in the region to the attention of the Colonial Department, and to ensure that the Barbadian agent in London would continue to supply the colony. MacGregor complained that although he had made repeated enquiries about vaccination since then, he had received no intelligence from the Board that the supply had been discontinued, 'the constant receipt and distribution of which, it is of so much importance to secure to the community'.[70] The seriousness of the Board's error soon became apparent when they could only secure a very small quantity of vaccine lymph from the military medical officer on the island. On 23 August, Butcher was forced to 'respectfully' advise the Governor that he needed to procure a supply 'from abroad'. Butcher recommended that a vessel be sent immediately to the neighbouring French colony of Martinique, where he 'confidently expected' that the island's authorities would 'reciprocate [MacGregor's] acknowledged kindness, by supplying our present necessities in this particular'.[71] On the 24 August, with all the humility that his request required, MacGregor wrote to the Governor of Martinique to 'Entreat Your Excellency's pardon for an intrusion reluctantly forced upon me by the extreme urgency of the case' and requested that Martinique spare a supply of vaccine lymph for the service of Barbados.[72]

A year later, the journal of the naval vessel *Cleopatra* reveals that Barbados was again short of lymph. The *Cleopatra* had been ordered to transport soldiers of the 76th Regiment from Bermuda to Barbados. Discovering cases of smallpox among the men, surgeon Martyn had, with difficulty, obtained some vaccine in both places but none 'was of any use'.[73] This time, the Board of Health was not to blame: Mr Mayers, the Barbadian Agent, complained to the Colonial Office that the vaccine lymph had 'in every instance failed', because it had been imperfectly packaged in Britain. Mayers was more surprised at this total failure, because he had personally applied for and received the supplies from both the Jennerian and the National Vaccine Establishment. In some instances he had even seen the very healthy and youthful English children from whose arms it had been extracted on the days immediately before the West India Mails left London. Mayers had taken some of the packages from the Jennerian Society directly to the Colonial Office and they 'appeared to be provided with due care and in the most secure packages'.[74] On 8 September 1840, John Gilham, the Inspector and Vaccinator for the National Vaccine Establishment, replied to the complaints from Barbados. He regretted the failure, but pointed out that 'the same source carefully packed in the same mode' had worked fine in the Channel Islands. He signed off by thanking Mayers for his 'zealous and unremitting personal and written applications', which had constantly 'advanced the *cause* in Barbados'.[75]

Conclusion

In the first decades of the nineteenth century, the movement of lymph became bound up in changing patterns of regional, transatlantic and global migration, and the uncertainties of a rapidly changing colonial situation. Surgeons' journals from emigrant ships to Canada in 1825 and colonial dispatches from Barbados in the 1830s show that getting vaccine to work as it travelled round the world was much more complicated than simply recruiting people to advance 'the *cause*'. As voyage after voyage failed to convey 'English' vaccine to Britain's colonies, there is little sense from the official record that its recipients asked questions about the ethnicity, race and class of the bodies through whom the vaccine matter had passed, as they would come to do later in the nineteenth century. Indeed, Sheridan notes that in Jamaica, colonists preferred locally obtained vaccine matter, rather than that of unknown origin sourced from London.[76] If this observation is noteworthy in the context of Irish migration or convict transportation, it is particularly remarkable given the extreme prejudice against colour that continued to exist in the Caribbean, and particularly in Barbados, at this time.[77] The extent to which ideas of race permeated discussions about vaccine in this earlier period is an important question and requires further research beyond the scope of this study. It is safe to say, however, that vaccine's early history lurks in the grey areas of freedom and compulsion that characterised a great deal of the history of colonial medicine. Considering the frustrations of making vaccine matter work in particular locations, when it was tried again and again, enables us to see that its supporters had to do more than simply use existing colonial networks to advance a 'cause'; they had to embed the practice in new colonial situations – to make vaccine's medical and colonial value mutually sustaining. The cases of Quebec and Barbados suggest that vaccination came to play a key role in the debates about immigration, freedom and labour that characterise this period.

Presenting packages of live vaccine conferred prestige and authority on those who controlled its distribution. Power relationships are thus critical to this story, because these gifts required human carriers to act as reservoirs of live matter, particularly across these interstitial maritime spaces of free and forced colonial migration. As they moved, different groups of people were seen to be highly vulnerable to disease, but it was precisely their vulnerability in marginal and beholden positions relative to those representing government that afforded medically and politically minded men ample opportunity to entangle vaccine in bigger colonial questions. As medical men tried to make vaccine work, Irish migrant children, convicts and African slaves emerge from the colonial record. It is by attending to the role of these people, as much as the self-professed imperial philanthropists, that we can see how vaccine's spread could be sustained.

Notes

1. Royal Jennerian Society, *Address of the Royal Jennerian Society, for the Extermination of Smallpox, with the Plan, Regulations, and Instructions for Vaccine Inoculation* (London: Royal Jennerian Society, 1803), 24–5.
2. Edward Jenner, *An Inquiry into the Causes and Effects of the Variolae Vaccine* (London: Sampson Low, 1798).
3. David Arnold, *Colonizing the Body* (London, Berkeley and Los Angeles: California University Press, 1993), 139–40.
4. Wellcome Library (WL), EPH+30:4, National Vaccine Establishment (London), 'Instructions Respecting Vaccination', 7 March 1809.
5. Michael Bennett, 'Smallpox and Cowpox under the Southern Cross: The Smallpox Epidemic of 1789 and the Advent of Vaccination in Colonial Australia', *Bulletin of the History of Medicine*, 83 (2009), 37–62, 51.
6. Martha Few, 'Circulating Smallpox Knowledge: Guatemalan Doctors, Maya Indians and Designing Spain's Smallpox Vaccination Expedition', *British Journal for the History of Science*, 43 (2010), 519–37; Catherine Mark and José G. Rigau-Pérez, 'The World's First Immunization Campaign: The Spanish Smallpox Vaccine Expedition, 1803–1813', *Bulletin of the History of Medicine*, 83 (2009), 63–94.
7. Mark Harrison, *Disease and the Modern World* (Cambridge: Polity, 2004), 64.

8. J.Z. Bowers, 'The Odyssey of Smallpox Vaccination', *Bulletin of the History of Medicine*, 55 (1981), 17–33; Niels Brimnes, 'Variolation, Vaccination and Popular Resistance in Early Colonial South India', *Medical History*, 48 (2004), 199–228. See also the essays in the Special Issue of the *Bulletin of the History of Medicine*, 83 (2009).

9. Andrea Rusnock, 'Catching Cowpox: The Early Spread of Smallpox Vaccination, 1798–1810', *Bulletin of the History of Medicine*, 83 (2009), 17–36, 22; Michael Bennett, 'Jenner's Ladies: Women and Vaccination in Early-Nineteenth Century Britain', *History*, 93:312 (2008), 497–513.

10. David Arnold, 'Smallpox and Colonial Medicine in Nineteenth-Century India', in David Arnold (ed.), *Imperial Medicine and Indigenous Societies: Disease, Medicine and Empire in the Nineteenth and Twentieth Centuries* (Manchester: Manchester University Press, 1998), 45–65.

11. Nadja Durbach, *Bodily Matters: The Anti-Vaccination Movement in England, 1853–1907* (Durham, NC and London: Duke University Press, 2005), 6; Arnold, 'Smallpox and Colonial Medicine', 57. Following Arnold's lead, vaccination in late nineteenth-century India has been particularly well investigated. See Sanjoy Bhattacharya, Mark Harrison and Michael Worboys, *Fractured States: Smallpox, Public Health and Vaccination Policy in British India, 1800–1947* (New Delhi: Orient Longman, 2005); Harish Naraindas, 'Care, Welfare and Treason: The Advent of Vaccination in the 19th Century', *Contributions to Indian Sociology*, 32 (1998), 67–96.

12. Melissa Leach and James Fairhead, *Vaccine Anxieties: Global Science, Child Health & Society* (London: Earthscan, 2007), 2.

13. Alison Bashford, *Imperial Hygiene. A Critical History of Colonialism, Nationalism and Public Health* (Houndmills: Palgrave Macmillan, 2004), 15, 25; Alison Bashford, 'Foreign Bodies: Vaccination, Contagion and Colonialism in the Nineteenth Century', in Alison Bashford and Claire Hooker (eds), *Contagion: Historical and Cultural Studies* (London and New York: Routledge, 2001), 39–60, 40.

14. Bashford, 'Foreign Bodies', 53–4.

15. Bowers, 'Odyssey', 17.

16. M.F. Wagstaffe, *Strictures on the Cow Pox, or Vaccine Inoculation* (Southwark: W. Kemmish, 1800), 1; John Coakley Lettsom, MD, *Expositions on the Inoculation of the Smallpox and of the Cow Pock*, 2nd edn (London: H. Fry, 1806), 1, 9, 11.

17. Thomas Trotter, *Medicina Nautica*, Vol. 1, 2nd edn (London: Longman, Hurst, Rees and Orme, 1804), 387–8.

18. Ibid., Vol. 2 (London: Longman, Hurst, Rees and Orme, 1799), 124–7.

19. Brian Vale and Griffith Edwards, *Physician to the Fleet: The Life and Times of Thomas Trotter, 1760–1832* (Woodbridge: Boydell Press, 2011), 137.

20. WL 20960/D, Papers Relating to Vaccination in Dublin, No.3 Report of the Cow-Pock Institution for 1810, 1.

21. See, for example, The National Archives (TNA) Admiralty Collections (ADM) 101/38/7, Journal of surgeon William Elyard on convict ship *John Bull* (1821–2), entries for 17 July, 24 July, 14 August 1821; TNA ADM 101/47/2, Journal of surgeon Matthew Anderson on convict ship *Mangles* (1822), General Remarks. This is important, because historians have previously assumed that there is 'no evidence' of serial inoculation of convicts being

put into practice: Bennett, 'Smallpox and Cowpox under the Southern Cross', 61.

22. Alan G. Brunger, 'Geographical Propinquity among Pre-Famine Catholic Irish settlers in Upper Canada', *Journal of Historical Geography*, 8 (1982), 265–82, 267–72. A smaller group of emigrants had already gone to Quebec in 1823, a plan designed both to alleviate the distress in Ireland caused by the potato harvest failure in 1821 and to provide recruits for the militia in Upper Canada. The emigrants were superintended by Peter Robinson, brother of the Upper Providence's Attorney-General. Marjory Harper and Stephen Constantine, *Migration and Empire* (Oxford: Oxford University Press, 2010), 17; Elizabeth Jane Errington, *Emigrant Worlds and Transatlantic Communities: Migration to Upper Canada in the First Half of the Nineteenth* Century (Montreal: McGill-Queen's University Press, 2007), 16, 29. For a detailed study of the Peter Robinson Settlers, see Carol Bennett, *Peter Robinson's Settlers* (Renfrew: Juniper Books, 1987).

23. TNA ADM 101/77/8, Journal of surgeon William Burnie on emigrant ship *John Barry* (22 April–25 July 1825), General Remarks. Before the arrival of Asiatic cholera, the term cholera was also used to describe a seasonal complaint.

24. TNA ADM 101/77/1, Journal of surgeon Francis Connin on the emigrant ship *Fortitude* (28 April–July 1825), ff.1-2; TNA ADM 101/76/9, Journal of surgeon P. Power on the emigrant ship *Elizabeth* (4 May–21 July 1825), Comments for 18 & 26 May, 8 June, f.2.

25. TNA ADM 101/76/3, Journal of James McTernan on the emigrant ship *Amity* (5 April–9 July 1825), General Remarks, ff.22-3.

26. Three years after its foundation, several prominent members of the Jennerian Society seceded and founded the London Vaccine Institution (LVI). By 1809 the original Jennerian Society had been dissolved, but was re-established in 1813 by the directors of the LVI. Thereafter, although the two institutions technically remained separate, both were registered at Burr St, and shared Director, John Walker, and Secretary, Dr Johnstone. 'Report of the Select Parliamentary Committee' reproduced in *London Medical Gazette*, Vol. XIII (Vol. 1 for 1833–4), 123–30, 125; Pigot & Co's *Metropolitan Guide & Book of Reference* (London: Pigot & Co., 1824), 142.

27. TNA ADM 101/76/2, Journal of surgeon John Thomson on the *Albion* (4 April–4 July 1825), conveying emigrants from Cork to Quebec. 'List of persons vaccinated' and General Remarks, ff.7, 22–3.

28. WL EPH+30:1, J. Creighton, Report laid before the Governors of the Foundling Hospital, Dublin, 4 January 1810, in *Annual report for 1809, of the Directors of the Cow-Pock Institution in Dublin* (Dublin: Cow-Pock Institution, 1810), 4; 'Foundling Hospital, Dublin', *Medical and Physical Journal*, 24 (July–December 1810), 522–3; 'Report of the Dublin Cow-Pock Institution', *Philosophical Magazine*, 36 (1810), 100–2.

29. Stephen Howe, *Ireland and Empire: Colonial Legacies in Irish History and Culture* (Oxford: Oxford University Press, 2000), xv.

30. On children as migrants, see Harper and Constantine, *Migration and Empire*, 247–76.

31. TNA ADM 101/13/8, Journal of surgeon John Tarn on emigrant ship *Brunswick* (5 April–27 June 1825), General Remarks, f.31.

32. TNA ADM 101/77/8, Journal of surgeon William Burnie on the emigrant ship *John Barry*, General Remarks, ff.2, 42–3.

33. TNA ADM 101/1/9, Journal of surgeon Thomas Logan on convict ship *Albion* (1828–9), General Remarks, f.30.

34. TNA ADM 101/76/2, Journal of surgeon John Thomson on *Albion* (4 April–4 July 1825), conveying emigrants from Cork to Quebec. 'List of persons vaccinated' and General Remarks, ff.7, 22–3.

35. Deborah Brunton, *The Politics of Vaccination: Practice and Policy in England, Wales, Ireland, and Scotland, 1800–1874* (Rochester, NY: University of Rochester Press, 2008), 15–16.

36. British Parliamentary Papers, 'Vaccine Establishment': Copy of Report received by the Secretary of State for the Home Department, 1825 [39] XV, 39–40.

37. Royal Jennerian Society, *Yearly Report* (London: John Westley and Co., 1830), 41.

38. John Thomson, *Historical Sketch of the Opinions Entertained by Medical Men Respecting the Varieties and the Secondary Occurrence of Small-pox* (London: Longman, Hurst, Rees, Orme and Brown/Edinburgh: David Brown, 1822), 397–8. This appears to be a different John Thomson than the emigrant surgeon discussed above.

39. John Jennings Cribb, *On Small-pox and Cow-pox* (Cambridge: W. & W. Hatfield, 1825), 70.

40. See, for example, Dr Erdmann, 'On Vaccine Matter', *Lancet*, 8:205 (4 August 1827), 553–4; John Leeson, 'Phenomena in Vaccination', *Lancet*, 12:303 (20 June 1829), 364; Richard Laming, 'On Vaccination', *Lancet*, 12:305 (4 July 1829), 420–1 and subsequent correspondence; William Howison, 'Remarks on Vaccination', *Lancet*, 16:411 (16 July 1831), 494–7.

41. Anon., 'Small-Pox Hospital', *Lancet*, 2:37 (12 June 1824), 349–51, 350; Dr. Delagrange, 'On the Present State of Vaccination in France', *Lancet*, 12:310 (8 August 1829), 582.

42. John Epps, MD, 'London Vaccine Institutions', *Lancet*, 16:406 (11 June 1831), 331–3.

43. TNA ADM 101/1/9, Logan, Journal on *Albion*, General Remarks, f.30.

44. King's College London, Special Collections: Foreign and Commonwealth Office Collection. Emigration Tracts Vol. XI, 1832–1840. 'Instructions for the surgeon's superintendent on board emigrant ships proceeding to New South Wales or Van Diemen's Land' (1838), Instruction 30.

45. British Parliamentary Papers, *Emigration*, 1839 [536–1] [536 II] XXXIX, 57.

46. Admiralty, *Instructions for Surgeons-Superintendents on Board Convict Ships* (London: William Clowes, 1838), Instruction 18.

47. 'P. Reilly, Esq. RN', *The Sydney Herald*, 8 July 1839, 2.

48. 'P. Reilly, Esq. RN', 2. For further discussion on emigrant and convict voyages to Australia, see Katherine Foxhall, *Health, Medicine and the Sea: Australian Voyages: c.1815–1860* (Manchester: Manchester University Press, 2012), Chapter 5.

49. See, for example, John Ring, *A Caution Against Vaccine Swindlers and Imposters* (London: J. Callow, 1816).

50. Epps, 'London Vaccine Institutions'; Medicus, 'Reply to the Statement of Dr Epps', *Lancet*, 16:412 (25 June 1831), 416.

51. For more on biomedical exchange, see Warwick Anderson, 'The Possession of Kuru: Medical Science and Biocolonial Exchange', *Comparative Studies in Society and History*, 42:4 (2000), 713–44.

52. The 1833 Act of Emancipation formally ended slavery in the British Empire on 1 August 1834. In the West Indies a transitional period known as apprenticeship existed until 'full' emancipation on 1 August 1838. See Melanie J. Newton, *The Children of Africa in the Colonies: Free People of Color in Barbados in the Age of Emancipation* (Baton Rouge, LA: Louisiana State University Press, 2008).

53. As Governor of the Windwards, MacGregor was responsible for receiving all correspondence from the Lieutenant Governors of the other islands and selecting the matters to be brought to Colonial Office attention. D.J. Murray, *The West Indies and the Development of Colonial Government 1801–1834* (Oxford: Clarendon Press, 1965), 181.

54. TNA Colonial Office (CO) 28/119, Barbados, Original Correspondence, Dispatches from Evan Murray John MacGregor, Governor of Barbados (January–July 1837). Letter to Lord Glenelg from Governor Sir Evan Murray MacGregor forwarding copies of correspondence respecting supplies of vaccine lymph (22 March). ff.158–61.

55. Richard B. Sheridan, *Doctors and Slaves: A Medical and Demographic History of Slavery in the British West Indies, 1680–1834* (Cambridge: Cambridge University Press, 1985), 249–53; Kenneth Kiple, *The Caribbean Slave: A Biological History* (Cambridge: Cambridge University Press, 1984), 144–5.

56. Larry Stewart, 'The Edge of Utility: Slaves and Smallpox in the Early Eighteenth Century', *Medical History*, 29 (1985), 54–70, 70.

57. B.W. Higman, *Slave Populations of the British Caribbean: 1807–1834* (Kingston: University of the West Indies Press, 1995) 278–9; Sheridan, *Doctors and Slaves*, 258–63.

58. TNA CO 28/119, Dispatches from MacGregor, L. Carmichael Smyth to MacGregor, 27 February 1837, f.159.

59. TNA CO 28/119, Dispatches from MacGregor, MacGregor to Glenelg, 1 May 1837, f.273.

60. Newton, *Children of Africa*, 4–7.

61. TNA CO 101/83, Grenada, Original Correspondence. MacGregor to Glenelg, 3 June 1837, ff.214–15.

62. On Mixed Commissions, see Farida Sheikh, 'Judicial Diplomacy: British Officials and the Mixed Commission Courts', in Keith Hamiltom and Patrick Salmon (eds), *Slavery, Diplomacy and Empire: Britain and the Suppression of the Slave Trade* (Eastbourne: Sussex Academic Press, 2009), 42–64, 43; Leslie Bethell, 'The Mixed Commission for the Suppression of the Transatlantic Slave Trade in the Nineteenth Century', *Journal of African History*, 7:1 (1966), 79–93.

63. TNA CO 101/83, MacGregor to Glenelg, 3 June 1837, f.214–6; TNA CO 101/84, Doyle to Glenelg, 25 September 1837, f.150.

64. TNA CO 101/83, Doyle to MacGregor, 21 May 1837, ff.218–9.

65. Newton, *The Children of Africa*, 10. See Letizia Gramaglia's chapter in this volume, for the impact of this form of migration on mental illness and institutionalisation in the region.

66. Bonham C. Richardson, 'Caribbean Migrations, 1838–1985', in Franklin A. Wright and Colin A. Palmer (eds), *The Modern Caribbean* (Chapel Hill and London: University of North Carolina Press, 1989), 203–28, 206; William A. Green, *British Slave Emancipation: The Sugar Colonies and the Great Experiment, 1830–1865* (Oxford: Clarendon Press, 1991), 261–71. The idea of a government-sponsored immigration from Sierra Leone drew strength from earlier experiments in assisted migration to Quebec, South Africa and Australia.
67. TNA CO 28/128, James Butcher, MD, President Board of Health to Joseph Garraway, Acting Private Secretary (Barbados), 17 August 1839, ff.61–2.
68. TNA CO 28/128, Garraway to Butcher, 18 August 1839, ff.62.
69. TNA CO 28/128, Butcher to Garraway, 19 August 1839, ff.62–3.
70. TNA CO 28/128, Garraway to Butcher, 22 August 1839, ff.64–5.
71. TNA CO 28/128, Butcher to Garraway, 23 August 1839, ff.66–7.
72. TNA CO 28/128, MacGregor to Governor of Martinique, 24 August 1839, f.67.
73. TNA ADM 101/94/2a, Journal of surgeon Patrick Martyn of the *Cleopatra* (1 February 1840–8 February 1841). Comments for June 1840 and General Remarks.
74. TNA CO 28/137, J.P. Mayers to James Stephen, Colonial Office, ff.18–9.
75. TNA CO 28/137, J.A. Gilham, National Vaccine Establishment to J.P. Mayers, Leg[islativ]e Agent, Barbados, 10 September 1840, ff.21–2.
76. Sheridan, *Doctors and Slaves*, 262. For late nineteenth-century debates, see Bashford, 'Foreign Bodies', 41.
77. Green, *British Slave Emancipation*, 17.

5
Victim or Vector? Tubercular Irish Nurses in England, 1930–1960

Anne Mac Lellan

> I didn't ever think I'd catch TB myself. I'd be of the type of Irish girl going over to England, working there, who didn't take much notice and did what I was told.[1]

In the decade following the founding of the new Irish state in 1922, a steady stream of young men and women migrated to England in search of work. Within ten years, this stream became a flood as England, with its open borders, replaced the newly restrictive United States as the destination of choice.[2] The areas in Ireland that experienced the greatest net migration were the 'least urbanized' and young emigrants leaving these areas had little exposure to tuberculosis (TB).[3] Irish rural migrants, many of them under 20 years of age, became an important part of the English labour force particularly in the healthcare sector. There was a severe shortage of nurses in England from the 1920s through to the 1950s and demand was further exacerbated during the Second World War.[4] Recruiting nurses to tend to tubercular patients was especially difficult.[5] Female Irish migrants helped fill this pressing need, but many of these young women were found to be themselves tubercular. Thus, in addition to presenting a need for care rather an ability to work, they seemed to pose a threat to their new country by acting as a source of infection. The belief that Irish migrants brought TB with them and were responsible for seeding it into English communities was common among members of the medical profession as well as the public.[6]

This chapter will examine evidence suggesting that between 1930 and 1960 Irish nurses were proportionately more tubercular than their English counterparts. But the construction of Irish nurses as carriers of TB into England was based on a misconception. It gradually became evident to the medical profession and policy makers that the converse was true, and that young nurses' lack of exposure to TB in Ireland rendered them vulnerable to infection soon after their arrival in England. As has been pointed out by John Welshman, the British government's attitude to the health status of Irish immigrants was determinedly low-key, with no compulsory medical examination at port of entry in the 1950s.[7] In Ireland, meanwhile, concerns were expressed about the health of young migrants in England, and the tendency for sick migrants to return home for care proved contentious. Returning tubercular Irish emigrants were often constructed by the Irish medical profession and in the Irish media as contributing to the country's already substantial TB problem. In 1957, it was claimed that 10 per cent of TB cases on the Irish register resulted from infection by returning emigrants.[8] Yet, the response of the Irish government to the issue of migrants and TB was just as low key as that of England. The Irish government prompted the National BCG Committee, established in 1949 under the aegis of Irish childhood TB specialist Dorothy Price, to increase awareness among intending emigrants of the usefulness of tuberculin testing and preventive BCG vaccination. Compulsory vaccination was mooted but rejected, and screening and vaccination remained voluntary.[9]

John Welshman and Alison Bashford's analysis of post-war debates about disease, borders and geographies of difference have shown that health policy was influenced by political, economic and employment imperatives. This chapter will test this analysis with respect to young Irish women migrating to England to train or work as nurses between 1930 and 1960. Nicholas King has pointed to the importance of essentialist versus non-essentialist explanations of disease in the context of immigration. Essentialist versus non-essentialist debates with respect to disease susceptibility in geographically or racially defined groups are somewhat similar to the nature versus nurture debates which focus on the relative contributions of genetics and environmental factors to the development of an individual. An essentialist understanding of susceptibility to TB focuses on pre-existing factors such as race, ethnicity and nationality, while a non-essentialist understanding focuses on contingent factors such as poverty and social disparity.[10] King has argued that political and practical responses may depend on the preferred understanding. This chapter will show that a complex and uneven understanding of Irish nurses' susceptibility to TB developed in England, and even when the misconception about their role as vectors of TB was dispelled, theories of racial susceptibility remained. Exploration of shifts in understanding and the experiences of tubercular Irish nurses between 1930 and

1960 will build on King's discussion of the importance of essentialist and non-essentialist explanations of disease and Welshman's arguments with respect to pragmatism and disease screening for Irish emigrants.

There was a sharp drop in TB mortality and morbidity in Ireland and England in the late 1950s. However, the disease remained endemic in many developing countries while developed countries continue to experience sporadic outbreaks. The chapter will seek to establish the relevance of the debates on Irish emigrant nurses that took place between 1930 and 1960 to current debates about the immigration of nurses to Ireland from developing countries. In 2000, Ireland responded to a shortage of nurses by actively recruiting overseas staff. Irish hospitals are now heavily reliant on non-EU migrant nurses.[11] Many of these nurses came from India and Pakistan where TB is endemic, and concerns about the possibility of these nurses bringing TB into Ireland have generated debate about the need for the introduction of screening measures for non-EU immigrant nurses taking up employment in Ireland.[12]

Irish nurses and recruitment issues in England

Between 1930 and 1960, there was a large increase in the number of nurses and nurse trainees employed in England and Wales. However, demand for nurses always outstripped supply.[13] In 1937, there were almost 40,000 registered nurses in English and Welsh hospitals with an additional 36,000 student nurses.[14] In 1945, the British government inaugurated a campaign to recruit more nurses.[15] Despite this, two years later, a shortfall of 30,000 nurses was reported.[16] By 1950, the numbers employed in hospitals in England and Wales had increased substantially to include almost 47,000 registered nurses, 16,600 enrolled nurses and more than 48,000 student nurses, and by 1962, there were 64,000 registered nurses, almost 14,000 enrolled nurses and 55,000 students staffing English and Welsh hospitals.[17] Irish trainees and qualified nurses were actively recruited through Irish labour exchanges and by British state authorities and hospitals.[18] Advertisements were placed in the local and national press and in nursing journals.[19] For young Irish women, nurse training provided the opportunity to enter a profession with 'high social status'.[20] More than 20,000 young Irish women took up nurse training or employment in Britain between 1940 and 1951.[21] An estimate of the proportion of Irish nurses in English hospitals and sanatoria may be gleaned from the Prophit Tuberculosis Survey, instigated by a commission appointed by the Royal College of Physicians in England in 1934. Its 'main object' was to determine whether it was possible to identify groups in society likely to develop TB. The members of the commission selected five groups for inclusion in the Survey: people with a case of active TB in their family (contacts), a control group, mainly office workers, thought to have a risk of exposure to TB similar to the 'average citizen', nurses and medical students who had an occupational risk of exposure to TB and entrants to naval training establishments who were young and lived in close proximity to each other. It was planned to observe 5,000 persons within each group. With the advent of the Second World War, however, the navy group had to be abandoned, while members of the other groups, with the exception of nurses, were scattered and difficult to track. In the final survey 10,000 adults, who were described as 'presumably healthy', were observed between 1934 and 1944.[22] Half of these adults were nurses and hospital nurses became a much more important part of the study than had been originally envisaged.

The hospitals in which these nurses worked were divided in the Survey into two categories or types. Type A hospitals had a 'heavy' working load with 30 to 43 nurses per 100 patients, while Type B hospitals with their 'light' working load had 67 to 72 nurses per 100 patients. Type A hospitals admitted general cases and the chronically ill, including advanced cases of various diseases (10–14 per cent of admissions died), while Type B hospitals typically did not admit chronically ill or advanced cases (6–8 per cent of admissions died). There were physical differences between the hospitals as well; for example, there was an average of six feet between beds in Type A hospitals and nine feet in Type B. The placing of beds close together meant a greater risk of infection and a more difficult working environment for nurses. Not surprisingly, Type A hospitals experienced more difficulties when recruiting staff and employed proportionately more Irish and Welsh than English nurses. Of the 3,046 nurses in Type B hospitals, only 4.9 per cent were Irish, while of the 1,969 nurses in Type A hospitals, 28.1 per cent were Irish. TB hospitals and sanatoria experienced the same difficulties as Type A general hospitals with respect to recruitment and it can be inferred that their staff also included high proportions of Irish girls who seemed prepared to work in poor conditions, and were, therefore, vital in times of overall nurse shortages. Hence, there were powerful pragmatic reasons for the Ministry of Health to leave the recruitment process open without introducing health-screening barriers. Elsewhere in this volume, Welshman has suggested that in the case of TB screening of migrants between

1950 and 1965 there were strong forces in favour of compulsory medical examination but that this did not come about.[23] The experience within the nursing sector fits well with Welshman's explanation of pragmatism driving policy.

Nurses and TB infection

Nurses, notably newly appointed staff, were exposed to more infectious diseases and were more prone to illness, including TB, than their peers in non-clinical settings.[24] However, TB was difficult to diagnose and its stigmatisation could lead to the concealment of symptoms.[25] Young Irish nurses may have been particularly sensitive to the popular notion that the Irish were inherently tubercular. The typical symptoms of pulmonary disease, the most common form of TB in an adult, include fever, night sweats, cough, loss of weight, difficulty in breathing and haemoptysis (spitting or coughing up blood). A tubercular patient might exhibit some but not all of these symptoms and patients suffering from different diseases such as pneumonia, typhoid or bronchial carcinoma could present a similar clinical picture.[26] X-rays and tuberculin skin tests were used as screening tools. A positive tuberculin test indicated an earlier infection or existing disease; in the case of the former, it implied a degree of immunity from TB. A person who was tuberculin negative had not been exposed to TB and, as such, was susceptible to infection.[27] Meanwhile, some infected persons never went on to develop the disease.[28] In addition to its use in individual diagnosis, medical practitioners conducted tuberculin testing of various populations to map the epidemiology of TB in England and Ireland between 1930 and 1960.[29] Histories of TB in Ireland in the twentieth century rarely refer to these epidemiological studies.[30] However, this chapter will demonstrate that tuberculin-testing surveys were of paramount importance when it came to understanding emigrant Irish nurses' experience of TB and the construction of them as carriers of the disease in England during these decades.

In the 1920s, concern about nurses' vulnerability to TB prompted a Norwegian study that demonstrated the usefulness of determining the tuberculin status of new nurse entrants. Johannes Heimbeck used tuberculin testing to determine previous exposure to TB among entrants to the nursing school at Ulevaal Hospital in Oslo.[31] He found that half of the student nurses were tuberculin negative at the time of entry. Virtually all of these tuberculin-negative nurses became infected during their three-year training. This was at variance with received wisdom that all young adults had been exposed to TB and were tuberculin positive. According to the former director of Norway's National Health Screening Service, Kjell Bjaarveit, the findings fundamentally 'changed the understanding of the pathogenesis of tuberculosis'. Student nurses were offered BCG vaccination by Heimbeck, while his colleague, Olaf Scheel, organised a similar project among medical students.[32] Heimbeck's work reduced the incidence of TB to one-sixth among the cohort of nurses who were vaccinated. Heimbeck found that 'once immunity, natural or BCG induced was established, there was very little subsequent morbidity'. The critical years for vaccination for nurses were during training when they were more likely to be tuberculin negative and vulnerable to infection. The positive effect of BCG seemed to be durable over the two decades of the study. In Scandinavian countries, tuberculin testing became widespread and mandatory for many occupational groups. Tuberculin testing was also espoused in the United States for various populations, including hospital staff. A study in Philadelphia in the 1930s found that 48 per cent of trainee nurses were tuberculin positive at entry and 100 per cent were tuberculin positive by the end of three years, implying that all trainees had come into contact with TB during training and had either developed the disease or immunity.[33]

The Prophit Survey in England relied on tuberculin testing as a screening test for susceptibility to infection and on x-rays for evidence of infection itself.[34] In England and Ireland, there was still some dispute as to the relative merits of x-rays and tuberculin testing.[35] As the Prophit Survey progressed, it became evident that the morbidity rates for nurses, and for medical students during their training on wards, was higher than for other occupational groups. Although the Survey was not designed to capture the experience of immigrant nurses, the additional dangers faced by these nurses soon emerged. Tubercular morbidity was two and a half times higher among Irish and Welsh nurses than their English counterparts. Indeed, the experience of Irish and Welsh nurses was so different that, for the purposes of overall conclusions, the Irish and Welsh cohorts were removed from the Survey to ensure that the results would not be skewed. The final report of the Prophit Survey, which commenced in 1943, was published in 1948 but a number of interim reports kept the issue under scrutiny as the Survey progressed. Another study carried out by Irish physician J.B. Lyons among the nursing staff based at Crumpsall Hospital in Manchester between 1943 and 1948 provided further evidence of the susceptibility of Irish nurses to TB. Lyons found that over the five years there were nine notified cases of TB with two fatalities.[36] Rural Irish girls accounted for five of the notified cases, including a fatal case of miliary disease, where tiny tubercular lesions of the size of millet seeds were disseminated throughout the body. Two further cases of TB were recorded among Welsh nurses and one case was English. The Irish nurses, who were a minority of the nursing staff in the hospital, contracted the most severe forms of TB. Lyons' explanation for Irish nurses' increased vulnerability to TB was based on a mixture of essentialist and non-essentialist understandings. He concluded that three factors determined Irish nurses' susceptibility to TB – 'a racial factor or what has been termed an absence of group immunity, an economic factor, and an absence of previous infection by the tubercule bacillus as judged by the absence of reaction to tuberculin'. Lyons did not explain the rationale behind his argument that Irish nurses' susceptibility to TB was a racial characteristic; perhaps, he felt it was self-evident. He explained that his definition of the 'economic factor' included nurses' 'yearly income, housing and place of employment'; however, in this article, he was referring solely to the latter. Lyons believed that 'the conditions under which any type of migratory Irish worker was employed were, as a rule, the most arduous obtaining for that type of work'.[37] The final report of the Prophit Survey recorded a morbidity rate of 27.1 per cent among the 984 Irish and Welsh nursing entrants and two of these died. The other 4,060 nursing entrants, who were mainly English, had a morbidity rate of 10.7 per cent. This was almost one-third that of the Irish and Welsh nurses, and, according to the final report, four of these died. Moreover, as in the Manchester study, the Irish and Welsh nurses were proportionately sicker than their English counterparts. Two-thirds of the tubercular Irish and Welsh nurses who took part in the Prophit Survey were treated in sanatoria compared with one-third of the tubercular English nurses. It was not possible to establish how tubercular Irish nurses who did not enter sanatoria were cared for, although it would appear that some of these nurses returned to their families in Ireland.

Preventive strategies

In the final report of the Survey it was concluded that the reason for higher rates of TB among Irish and Welsh nurses, given their 'equality of exposure and equality of environment', had to be due to a 'valid racial difference' between these Celtic nurses and their English peers.[38]

It was suggested that this racial difference might not be permanent and that group immunity might be produced by the survival of the least susceptible but 'by what process such genetic factors operate it is difficult to say. It seems possible that the inherited factor is not precisely one of immunity, but rather of native capacity to develop specific immunity as a result of an infection'.[39] This somewhat complex essentialist understanding of Irish and Welsh nurses' susceptibility to TB prompted the suggestion that Irish and Welsh nurses' risk of contracting TB might be reduced by 'a greater concentration on living and working conditions and exposure factors'. Regular tuberculin testing was proposed for all tuberculin-negative nurses irrespective of race.[40] It was recommended that tuberculin-negative nurses should not work in the TB wards of general hospitals while patients were to be 'well trained in the habits of coughing and expectorating' in order to reduce the possibility of infecting nurses.[41] The Survey did not recommend preventive vaccination for nurses but did suggest that vaccination might 'warrant further serious study'. In cases when a nurse became tuberculin positive, then regular x-rays were recommended. During the 1940s, hospitals in England instigated various regimes of surveillance and care for nurse entrants. These were post-employment measures, which were extended to all nurse entrants rather than singling out immigrants. In 1945, the General Nursing Council for England and Wales made it a requirement for hospitals approved as nurse training schools to x-ray student nurses before or upon entry and to repeat the x-rays at intervals of a year or less.[42] A leading article in the *British Medical Journal* (*BMJ*) in 1948, published after the release of the final Prophit Survey report, claimed that 'genetic immunity clearly played a part' and amplified the suggestions of the Prophit Survey by proposing that 'all persons of 18 years of age or over should not only be kept under general medical surveillance but should be tuberculin tested periodically and, when necessary, x-rayed'.[43] R.W. Parnell, the physician in charge of the Student Health Service, Oxford University, pointed out the impracticality of such a strategy and suggested focusing on groups such as nurses where a 'special recognition of exposure is recognized'.[44] In 1950, nurses 'in close and frequent contact with tuberculosis' were insured against the disease under the National Insurance (Industrial Injuries) Act, 1946.[45]

In Ireland, there were various theories among members of the medical profession about the link between race and susceptibility to TB.[46] In 1939, Dorothy Price had published her findings that Irish adolescents had a surprisingly low level of tuberculin positivity and that there was a strong rural/urban divide; country dwellers were less likely to have been exposed to TB.[47] Price's results were similar to those obtained in the Prophit Survey which had found that tuberculin sensitivity among Irish and Welsh nurse participants in the Survey varied according to geographical origin; nurses from rural backgrounds were more likely to be tuberculin negative. Other studies in Ireland provided further evidence that tuberculin-negative nurses working in Ireland were 'particularly likely to suffer ill effects from tuberculous infection'.[48] Price and the Prophit Survey, however, interpreted the data differently. Price posited that Irish nurses were tuberculin negative due to a lack of previous exposure to disease rather than genetic factors. She proposed the introduction of the preventive vaccine, BCG. According to Price, with 70 to 80 per cent of young Irish adolescents 'not yet infected by the tubercle bacillus', it was important to conduct tuberculin test and vaccinate vulnerable groups such as young Irish nurses emigrating to England.[49]

Hospital authorities anxious to recruit nurses regarded highlighting the increased risks that nurses faced of contracting TB as counterproductive. For example, in 1945 when Alec Wingfield, physician to the Seaman's Hospital, London, suggested that the dangers of nursing the tubercular should be made more widely known, he was tersely rebuked by the medical superintendent and deputy medical superintendent of Cheshire Joint Sanitoria, Peter Edwards and A. Clark Penman. They argued that 'narrow publicity among those responsible for the health of hospital staff would be more useful and less likely to diminish even further the number of entrants to tuberculosis nursing'.[50] Edwards pointed out that 25 per cent of the student nurses in Cheshire Sanatorium were tuberculin negative on entry. If they were excluded, then a proportionate number of beds would be lost. In a subsequent paper on TB among sanatoria nurses, Edwards elaborated on his argument in support of employing tuberculin negative nurses. He suggested that a regime of care for tuberculin-negative nurses should be put in place and this should include frequent tuberculin tests and x-rays.[51] Edwards' study also pointed to the experiences of Irish emigrant nurses. He found a much higher incidence of tuberculin-negative nurses among this group. Of the 94 healthy Irish entrants, 45 per cent were tuberculin negative. These accounted for two-thirds of the overall tuberculin-negative cohort. Among the case studies detailed in his report, Edwards included an 18-year-old Irish girl who joined the staff on 19 September 1937. Case 12 was tuberculin negative but became positive in three months indicating she was recently infected:

Nine months later, she developed an acute febrile illness after a stormy crossing from Ireland while suffering from a heavy cold. A left pleural effusion developed which cleared rapidly without aspiration. She was in bed for three months and had a further month of modified duty. She is well and working as a ward sister in 1945.[52]

Edwards concluded that nurses who tested negative for tuberculin should be employed and looked after as it was 'not possible for the average country institution drawing its nurses and maids from rural districts and from Ireland' to exclude job applicants who were tuberculin negative.[53]

It is noticeable that the concern expressed about the vulnerability of Irish trainee nurses to TB was confined to doctors, often publishing in medical journals, which had a target audience of other doctors. Occasionally, Irish nursing journals published the views of doctors with respect to occupational TB but it was rare for nurses themselves to express worries about contracting the disease. It would seem that at this time, for nurses, anxiety about professionalisation took precedence over anxiety about occupational health. In 1946, B.M. Dunlevy, a medical officer employed by Dublin City Corporation's TB service, observed in the *Irish Nurses' Magazine* that 'when a girl enters the nursing profession she seldom thinks of the danger of infection from her patients yet she may debate the question of risk before taking up nursing in a sanatorium'.[54] This was certainly true in the case of Bridie G., an Irish nurse probationer in the London Jewish Hospital from 1940 to 1943. She contended that her attitude was typical of many young Irish girls, who were abroad for the first time, alone and in their teens. 'I'd be of the type of Irish girl going over to England, working there, who didn't take much notice and did what I was told'.[55] However, she 'was very aware' of TB in her home community in Ireland. 'They said it used to be transmitted through clothes or shoes... a pair of shoes handed down to a smaller one... the perspiration in the shoes. Anyone who had it, really it killed them'. Although she was afraid of contracting TB from neighbours in her Irish community, she did not consider the disease was a risk in her work as a nurse. 'I never worried about picking it [TB] up in a hospital'. Bridie had a chest x-ray when she arrived at the hospital but she does not recall being 'inoculated'.[56]

It is unsurprising that Bridie was not offered vaccination in England in 1940. The efficacy and safety of BCG, first used in humans in 1921, was widely debated throughout the first half of the twentieth century.

For reasons that seemed to have more to do with nationalism and local styles of medicine rather than scientific evidence, BCG vaccine was not made generally available to nurses in Ireland and England until the late 1940s.[57] BCG vaccine was the subject of competing discourses in the British medical journals throughout the mid- to late 1940s, with an increasing number of doctors favouring vaccination. An editorial in *Tubercule*, the journal of the British Tuberculosis Association, in September 1946 stated that 'in Mantoux [tuberculin] negative nurses, opinion appears unanimous that there is an obligation to use BCG'.[58] That same year, F.R.G. Heaf, London's Chief Tuberculosis Officer, and a supporter of BCG vaccination, wrote in the *Irish Journal of Medical Science* that a number of authorities were preventing tuberculin-negative nurses from working in TB wards until six months after conversion to a positive reaction.[59] The *British Journal of Nursing* reported on a deputation from the Tuberculosis Association to the Ministry of Health. This deputation had informed the Ministry that BCG was now regarded in Norway as 'so valuable that a tuberculin-negative nurse who has not been offered vaccination and who later developed tuberculosis might well have a claim for damages'.[60] F.B. Smith argues that BCG 'slipped into Britain about 1947 as propaganda to comfort sanatoria nurses' when there was a near collapse of the sanatoria due to the lack of staff.[61] However, BCG was not offered universally to nursing entrants and in January 1949, a leading article in *Tubercle* again decried the fact that nurses in England were 'still denied the use of this preventive measure'.[62] Later that year, the *British Journal of Nursing* reported that BCG vaccination would be offered in 'due course' to all hospital nurses and medical students. It was acknowledged that this would be a 'big task' but it would be 'steadily proceeded' with as quickly as the 'work could be done'.[63]

In Ireland, pre-emigration vaccination of nurses was also a big task that proceeded slowly. In 1951, two and a half years after the National BCG Committee and a separate Dublin Corporation scheme under the direction of B.M. Dunlevy had began work on mass vaccination, many young Irish girls going to England to train as nurses still had not been vaccinated. In 1951, J.B. Lyons stated that he had yet to meet a single migratory Irish worker who had had BCG vaccination.[64] In a study of 67 Irish nurses in training between 1943 and 1948, he found that 43 per cent were tuberculin negative prior to starting work on the wards. This meant that almost half of the Irish trainees had not come into contact with TB. Lyons wrote emotively of 'the absence of previous infection in those who had led comparatively sheltered lives or came from rural districts, when placed in a milieu where it is commonplace to encounter the infecting organism' and as young nursing trainees 'newly embarked upon their working careers' had 'been recently freed from parental restraint and deprived of maternal vigilance at a time of very active growth and endocrine change'.[65] Lyons did not mention other possible sources of tubercular infection such as nurses' homes where nurse probationers were usually accommodated.

The Irish National BCG Committee was acutely aware of the need to vaccinate emigrants before departure from Ireland. Indeed, as it expanded its work, it began to provide further evidence of this need for vaccination. Tuberculin testing was a precursor to BCG vaccination, so, as a by-product of its work, the National BCG Committee began to accumulate further statistics with respect to the tuberculin status of young people in various Irish counties. In 1952, an average of 50 per cent of young adults in six rural counties, in the age bracket of between 18 and 30 years, were tuberculin negative. The young people from these rural counties were the particularly likely to emigrate. In 1954, the County Medical Officer of Health for Roscommon, Michael Flynn, found that less than 15 per cent of young people in the county were tuberculin positive. He concluded that it was obvious that young adults who were potential emigrants were particularly vulnerable and the BCG campaign should concentrate on them.[66] The urgent need to reach these young adults and the difficulties in persuading them to volunteer for vaccination was a recurring theme in the annual reports of the National BCG Committee from 1949 to 1954. In 1951, an article in the *Irish Times* reported that it was no longer inevitable for the 'apparently strong country girl, studying nursing' to become 'delicate' with TB, which had recently 'been recognised as an occupational hazard among nurses'.[67] In 1954, the National BCG Committee's reported with evident pride that its findings had caused a 'complete reversal of medical opinion' concerning the young Irish emigrant and TB:

> It has, at last, been generally recognised that he is not to be shunned for fear of contracting tuberculosis from him, but rather in order to protect him from contracting the disease in an urbanised environment to which he has come, uninfected and unprotected.[68]

But, this change in understanding from a perception that Irish nurses' susceptibility to TB was due to genetic factors to a conviction that it was caused by a lack of prior exposure to the disease did not lead to increased vaccination rates among Irish nurse emigrants prior to departure from Ireland. In 1954, the report of the first Irish National Tuberculosis Survey criticised the slow rollout of the BCG campaign. In response, the *Irish Press* emphasised 'five facts on emigration', pointing out that a large proportion of Irish people emigrating from rural areas were tuberculin negative.[69] The *Irish Times* also reported on the number of emigrants who returned each year to Ireland as invalids, and the 'incalculable damage' done by these returning emigrants in infecting their relatives, not to mention the costs placed on the state.[70] The Government Information Bureau, on behalf of the Department of Health, asked Price to issue a summary of the situation and to encourage emigrants to seek BCG before going abroad.[71] Price suggested that compulsory BCG vaccination for every Irish emigrant should be considered. In England, Member of Parliament Dr Barnett Stross suggested that chest radiography and tuberculin testing of those intending to migrate, followed by BCG vaccination of those who were tuberculin negative, should be carried out by the Irish authorities, in order to protect young Irish people emigrating to England.[72] The National BCG Committee placed advertisements in Irish newspapers informing would-be emigrants that if they intended to emigrate, 'BCG vaccination against tuberculosis is as important for you as your travel ticket!' It also 'seemed expedient' to point out to young emigrants that if they did not get vaccinated at home, they could now do so in England, which had begun to make BCG freely available in 1956 following the first report of the British Medical Research Council's large-scale vaccine trial.[73] In December 1957, the National BCG Committee in Ireland commissioned posters in the English and Irish languages and these were displayed in 2,800 post offices in Ireland. Meanwhile, nursing schools at home and abroad were increasingly requesting records of vaccination.

In spite of the growing consensus with respect to the need for BCG vaccination for young Irish emigrants, Brigid E. who trained as a nurse in Rushgreen Hospital, Romford, Essex, between 1954 and 1957 reported that she did not receive a medical examination nor was she x-rayed. She began work at the age of 18 and stated: 'I was not offered a tuberculin test and was not offered BCG. It was very scary because there were some Irish nurses who contracted the disease. They were just sent back, very quietly, to Ireland. I was terrified'.[74] Brigid suggested that precautions against the transmission of infection were limited by 'a great shortage of nurses particularly at night'. Gloves were changed and hands washed but she says that there was 'no mention of changing gowns between patients'. Brigid was probably more aware of the dangers of contracting TB than most other probationers. In 1947, when Brigid was 11 years old, her 24-year-old sister contracted TB while working in a psychiatric hospital in England and died. It was suggested by the hospital that Brigid's sister contracted TB from feeding prisoners in a nearby prisoner-of-war camp. Brigid claimed that this was unlikely as the prisoners were behind wire netting and close contact was not possible. She described it as being 'like feeding the birds'. Brigid said that she was still distressed and that, for her, it was a 'cover-up' by the hospital authorities. Her sister died in England despite the efforts of her parents who had purchased streptomycin from America 'at great cost' to treat her; the drug, however, arrived too late. Throughout her training, Brigid says she worried all the time about contracting TB. Brigid said that there was still a 'stigma about tuberculosis similar to today's stigma with respect to mental illness'.[75]

During the mid- to late 1950s, as TB declined more rapidly in England than in Ireland, concern about tubercular Irish migrants extended way beyond the nursing community. J.B. Lyons' statement of 1951 that it was 'an axiom in English medical schools' to suspect TB in young Irish adults retained its relevance throughout much of the 1950s.[76] In 1954, Evelyn Hess and Norman McDonald, physicians at Clare Hall Hospital, London, published the results of a comparative survey of Irish patients and 'Londoners' born in the Greater London area.[77] The survey was carried out in five major hospitals in the North West metropolitan region of London. Hess and McDonald repeated a medical practitioner's comment that 'They bring it over with them. You will find whole families in Ireland who are rotten with it', while a TB officer was quoted as saying 'with obvious sincerity' that 'it's the Irish who are our greatest problem'. Hess and McDonald found that the ratio of tubercular Irish patients to Londoners was twice what might have been expected. They opined that obtaining a BCG vaccination before leaving Ireland should be considered for those migrants who were 'the immunologically ill-equipped descendent of rural stock, often in the tuberculin negative state and in the susceptible 15–25-year-old age bracket'.[78] In 1956 G.Z. Brett of the Mass Radiography Service, North West Regional Hospital Board, London, published a study of 32,000 persons examined by the service. He found there was an excess incidence of TB of at least three, and probably nearer seven, times among the Irish cohort compared to a control group. His conclusion was that in the face of declining TB mortality in England, the high incidence of active, infectious disease in Irish-born residents in England was of epidemiological significance.[79] Brett found extensive disease in 50 per cent of Irish male cases and 20 per cent of Irish female cases. He pointed out that this may have been an underestimate of the extent of female cases, as hospital staff members were not included in the study. In 1958,

V.H. Springett, a TB researcher who had worked for the Prophit Survey, extended the analysis of Irish immigrant experience of TB from London to Birmingham and found that notifications of TB in Irish-born people were twice as high as notifications among the English population.[80]

At this stage, Springett's conclusion that the excess diagnosis of the disease among Irish-born migrants was probably due to the migration of uninfected young adults into a relatively infectious environment echoed the prevailing paradigm among the medical profession in England and Ireland in the 1950s. Criticising the slow pace of the rollout of BCG vaccination in Ireland, he again suggested that it was appropriate prior to migration. The chairman of the British Joint Tuberculosis Council elaborated on these criticisms, arguing in a letter to the *BMJ* that a satisfactory vaccination scheme for young Irish immigrants had 'not yet been evolved'.[81] An analysis of the vaccinations carried out under the auspices of the National BCG Committee in Ireland demonstrates the validity of the various criticisms. For the first seven years of the programme (1949–1956), the young adult age group (15–29 years) consistently accounted for a very low proportion of the total vaccinations performed. Out of 300,000 vaccinated by 31 December 1956, fewer than 30,000 were young adults. This would suggest that approximately 5 per cent of the 640,000 young adult population aged between 15 and 29 years were vaccinated. This was the age group most likely to emigrate. In 1959, Dunlevy wrote that the 'deplorable situation' of migrants leaving Ireland without vaccination was 'unnecessary', as BCG was available throughout Ireland while in Dublin city a special clinic had been established in 1957 for the protection of young adults. This clinic was held twice weekly after working hours but the response was 'most disappointing'. Using the by-now familiar rhetoric, she added that

> Migration of the immunologically ill-equipped descendant of rural stock, often in the tuberculin-negative state and in the susceptible 15–25 age-group, to the great centres of population, where the risks of heavy infection are greater and the strain imposed by living conditions is much intensified, is considered to be one of the chief means whereby tuberculosis perpetuates itself as a social disease.[82]

Part of the explanation for the poor uptake of the vaccine may lie in the nature of tuberculin testing and BCG vaccination, which could require up to six visits to a doctor or clinic. In the end, it was only by a process of attrition that young Irish emigrants were protected against TB; by the early 1960s, it was reckoned that many of those in the emigration age bracket had been vaccinated during their school years under a scheme introduced by the National BCG Committee in the 1950s. By the late 1950s, endemic TB had effectively been wiped out in both Ireland and England. From the mid- to late 1950s, British concern about TB, immigrants and ethnicity extended beyond the Irish to other nationalities, in particular Indians, Pakistanis and West Indians who, in contrast to the largely tuberculin negative Irish, came from countries where TB was endemic.

Conclusion

Epidemiological surveys throughout the 1930s and 1940s demonstrated a rate of tubercular morbidity and mortality among Irish nurses in England that exceeded the rate among their English peers. Despite this evidence, it would seem that the pressing shortage of nurses provided a compelling reason to allow continued unimpeded recruitment of Irish nurses by English hospitals. This study of one specific professional group echoes and expands on the findings of Welshman and Bashford about the pragmatism displayed by the British Ministry of Health towards TB screening and Irish emigrants as a whole. Surveys carried out in Ireland and England, using tuberculin testing and x-rays for diagnosis, provided evidence of Irish emigrant nurses' lack of prior exposure to TB. These surveys gradually changed perceptions among the English and Irish medical professions and policy makers between 1930 and 1960 with respect to Irish nurses and their experience of TB in England. This finding augments Anne Hardy's demonstration of a general reframing of TB infection in the English population between 1938 and 1970 following the publication of large-scale epidemiological surveys which relied on tuberculin screening and x-ray in the diagnosis and tracing of TB infection. According to Hardy, the new techniques replaced the old 'romanticized image' of TB with one 'centred on children and the old'.[83] In the case of Irish emigrant nurses, their image as vectors of TB was replaced by a new representation as victims of the disease.

The events surrounding the assessment of tubercular Irish nurses indicate that essentialist and non-essentialist readings of TB susceptibility among young Irish nurses produced different policy proposals. The Prophit Survey recommended the surveillance of tuberculin negative nurses, and an improvement in working and living conditions, as the appropriate response to their contention that there was a genetic basis for susceptibility to TB among Irish nurses. In Ireland, it was suggested that it was the absence of contact with TB rather than genetic factors that accounted for Irish nurses' vulnerability to infection. As this theory gained acceptance in both Ireland and England, the medical profession in both countries advocated the introduction of BCG vaccination for tuberculin negative nurses. Welshman is critical of the reliance on biomedicine to provide protection against TB infection for immigrants. He has posited that the theory of susceptibility caused by limited contact with the disease deflected attention from the economic and social conditions encountered by Irish migrants in English cities. He has asserted that 'focusing on surveillance and biomedical factors' meant that it was possible to avoid confronting 'more radical political and environmental change'.[84] While this may be true in a broad context, it does not hold up in the case of immigrant Irish nurses' experience of TB. The Prophit Survey demonstrated that immigrant Irish and Welsh nurses who worked in similar circumstances to English nurses had divergent rates of TB. The arduous working conditions for nurses in Type A hospitals were not exclusive to immigrant nurses. English nurses, working and living in the same environment, had lower rates of TB. Hence, in the context of Irish nurses and trainee nurses in England, surveillance and biomedicine in the form of tuberculin testing and BCG vaccination provided the possibility of an efficient means of protection against TB infection in environments where susceptible individuals were exposed to high doses of the infecting organism.

Screening and preventive measures against TB were eventually inaugurated for all nurses – immigrant nurses were not singled out – in England at the post-employment rather than the recruitment stage. This chimes with Welshman's findings that pragmatism drove policy. In Ireland, pre-emigration publicity campaigns were instituted and screening and preventive vaccine were offered on a voluntary basis. The subsequent low uptake of BCG vaccine by intending emigrants was criticised in England and Ireland as a systems failure or, more specifically, a failure of the Irish National BCG Committee, rather than the failure of individual nurses to take charge of their own health.

Finally, in a reversal of the situation from 1930 to 1960, TB in Ireland is now perceived as a disease of immigration rather than emigration. More than 15,000 non-EU nurses presently work in Irish hospitals. Current Irish guidelines on the prevention and control of TB recommend that health-care workers arriving in Ireland or returning to Ireland from countries with a high incidence of TB should have a chest x-ray and a tuberculin test although it is not compulsory.[85] In 2010, discrepancies between the results of chest x-rays of Philippino and Indian nurses, taken in their country of origin, and later repeat x-rays taken in Ireland prompted a call in the *Irish Medical Journal* for a national TB screening programme to be instituted in Ireland for overseas nurses recruited to work in the country.[86] The concerns expressed about the difficulties of ensuring adequate standards of medical testing and x-ray film interpretation in the country of origin are remarkably similar to the concerns raised in the late 1950s in England. The perceived difficulties in policing emigration, as articulated by John Welshman, that led, in part, to pragmatic British policies with respect to TB screening in the 1950s are now being played out in Ireland.[87] More than half a century after Irish immigrants and TB were emotive topics in England, the White Plague is increasingly being perceived in Ireland as a disease of 'aliens'.

Notes

1. Interview, Mrs Bridie G., County Mayo, Ireland, 30 November 2011. Mrs G. trained as a nurse in the London Jewish Hospital between 1940 and 1943. The author would like to thank Bridie Gough for her permission to publish extracts from her interview.
2. Michael P. Flynn and J. Cyril Joyce, 'Migration and Tuberculosis: The Irish Aspect', *Tubercule*, 36:11 (1955), 336–44; Enda Delaney, *Demography, State and Society* (Liverpool: Liverpool University Press, 2000), 84.
3. Flynn and Joyce, 'Migration and Tuberculosis', 336.
4. Brian Abel-Smith, *A History of the Nursing Profession* (London: Heinemann, 1975), 115–17, 176–9.
5. Monica E. Baly, *Nursing & Social Change*, 3rd edn (London: Routledge, 1995), 170–1; Anon., 'Serviceman and Tuberculosis', *British Journal of Nursing (BJN)*, 93:2124 (1945), 80. According to Dr H.M. McAuley, the 'whole fabric of the anti-tuberculosis scheme is threatening to collapse' due to the lack of nurses and domestic staff.
6. John Haskey, 'Mortality among Second Generation Irish in England and Wales', *British Medical Journal (BMJ)*, 312:7043 (1996), 1373–4. According to Haskey, a statistician with the Census, Population and Health Group, Office for National Statistics, 'mortality of first-generation Irish people exceeds that of all residents of England and Wales by about 30 per cent for men and 20 per cent for women'.
7. See John Welshman, 'Tuberculosis, "Race", and Migration, 1950–70', *Medical Historian: Bulletin of Liverpool Medical History Society*, 15 (2003–4), 36–53; John Welshman, 'Compulsion, Localism, and Pragmatism: the Micro-Politics of Tuberculosis Screening in the United Kingdom, 1950–1965', *Social History of Medicine*, 19 (2006), 295–312; John Welshman and Alison Bashford, 'Tuberculosis, Migration, and Medical Examination: Lessons from History', *Journal of Epidemiology and Community Health*, 60 (2006), 82–4; John Welshman, 'Importation, Deprivation, and Susceptibility: Tuberculosis Narratives in Postwar Britain', in Flurin Condrau and Michael Worboys (eds), *Tuberculosis Then and Now: Perspectives on the History of an Infectious Disease* (Montreal and Quebec: McGill-Queen's University Press, 2010), 123–47; I. Convery,

John Welshman, and Alison Bashford, 'Where is the Border? Screening for Tuberculosis in the United Kingdom and Australia, 1950–2000', in Alison Bashford (ed.), *Medicine at the Border: Disease, Globalization and Security, 1850 to the Present* (Houndmills: Palgrave Macmillan, 2006), 97–115.

8. Leading article, 'Tuberculosis in Immigrants', *Tubercule*, 38:3 (1957), 217–9; Anon., 'BCG can Protect Emigrants', *Irish Times*, 27 October 1954. It was stated in *Tubercle* that many minor epidemics may have been created among 'tuberculin-negative near-relatives and others in Éire by bacilli imported from Great Britain'.

9. Greta Jones, *'Captain of all these Men of Death'. The History of Tuberculosis in Nineteenth and Twentieth Century Ireland* (Amsterdam and New York: Rodopi, 2001), 222.

10. Nicholas King, 'TB, Immigration and Race', in Matthew Gandy and Alimuddin Zumla (eds), *The Return of the White Plague* (London and New York: Verso, 2003), 44.

11. Niamh Humphries, Ruairi Brugha and Hannah McGee, 'Nurse Migration and Health Workforce Planning in the Irish Context', Irish Forum for Global Health 2012, http://www.globalhealth.ie/index.php?i=305 [accessed 10 July 2012]. Almost 15,000 non-EU migrant nurses were registered in Ireland between 2000 and 2010 accounting for 35 per cent of new nurse registrations.

12. S. Power, J. Sim, J. Gallagher and B. Greiner, 'A Study to Compare Chest X-ray Reports on Overseas Nursing Recruits', *Irish Medical Journal (IMJ)*, 103:5 (2010), 140–1.

13. Abel-Smith, *A History of the Nursing Profession*, 209–17.

14. Baly, *Nursing and Social Change*, 'Nursing Staff in Hospitals (England and Wales)', Table 18.1, 220.

15. Anon., *Staffing the Hospitals: An Urgent National Need*, HM Stationary Office (1945). Cited in 'Hospital Nurses: A New Deal', *BMJ*, 2:4429 (1945), 733–4.

16. Bethina A. Bennett, 'Nursing in Britain', *American Journal of Nursing*, 47:5 (1947), 281–4.

17. Baly, *Nursing and Social Change*, 'Nursing Staff in Hospitals (England and Wales)', Table 18.1, 220.

18. Nicola Yeates, 'Migration and Nursing in Ireland: An Internationalist History', *Translocations: Migration and Social Change e-journal*, 10 (2009), http://www.translocations.ie/docs/v05i01/Vol_5_Issue_1_d.pdf [accessed 10 December 2011].

19. The number of advertisements seeking nurses and trainee nurses for British hospitals consistently outnumbered the number of vacancies advertised in Irish hospitals in *Irish Nursing and Hospital World* between 1940 and 1952: Yeates, 'Migration and Nursing in Ireland', 10.

20. Nicola Yeates, 'A Dialogue with "Global Care Chain" Analysis: Nurse Migration in the Irish Context', *Feminist Review*, 77 (2004), 79–95.

21. Yeates, 'Migration and Nursing in Ireland'; Delaney, *Demography, State and Society*, 135; Eleanor C. Grogan, 'The Nurses' Exodus', *Irish Nurses' Magazine (INM)*, 12:28 (1943), 1–2. Grogan complained that so many nurses were going to Britain that 'for the first time since trained nursing became a professional career Ireland faces a shortage of nurses'.

22. Marc Daniels, Frank Ridehalgh, and V.H. Springett, *Tuberculosis in Young Adults. Report of the Prophit Tuberculosis Survey 1935–1944* (London: H.K. Lewis, 1948), v. The Survey was funded by a legacy that J.M.G. Prophit left to the Royal College of Physicians in order to carry out research into TB which was 'the main killing and incapacitating disease affecting the productive groups of the population'. A series of scholars were appointed in succession: Frank Ridehalgh, I.M. Hall, Mark Daniels and V.H. Springett.
23. See John Welshman's chapter in this volume.
24. Anon., 'Hospital Infections', *BJN*, 85:2030 (1937), 250; Anon., 'Tuberculosis among Student Nurses', *BJN*, 90:2084 (1942), 52.
25. *Department of Local Government and Public Health Report 1936–1937* (Dublin: Government Stationary Office, 1938), 174.
26. Linda Bryder, *Below the Magic Mountain. A Social History of Tuberculosis in Twentieth-Century Britain* (Oxford: Clarendon Press, 1988), 4–5.
27. Daniels, Ridehalgh and Springett, *Prophit Survey*, 10.
28. Health Protection Surveillance Centre, *Tuberculosis Guidelines 2010* (Dublin: Health Protection Surveillance Centre, 2010), 190.
29. Anne Hardy, 'Reframing Disease: Changing Perceptions of Tuberculosis in England and Wales, 1938–70', *Historical Research*, 76 (2003), 535–56. Tuberculin surveys became meaningless once the use of BCG vaccination became widespread as BCG renders a person tuberculin positive.
30. Margaret Ó hÓgartaigh, 'Dr Dorothy Price and the Elimination of Childhood Tuberculosis', in Joost Augusteijn (ed.), *Ireland in the 1930s* (Dublin: Four Courts Press, 1999), 67–82; Jones, *'Captain of all these Men of Death'*, 145, 157. Ó hÓgartaigh recognised the utility of the tuberculin test with respect to testing children in Peamount Sanatorium in the late 1930s. Jones briefly mentions tuberculin studies carried out by Irish physician Dorothy Price.
31. K. Bjartveit, Olaf Scheel and Johannes Heimbeck, 'Their Contribution to Understanding the Pathogenesis and Prevention of Tuberculosis', *International Journal of Tuberculosis and Lung Disease*, 7:4 (2003), 306–11; Johannes Heimbeck, 'BCG Vaccination of Nurses', *Tubercule*, 29:4 (1948), 84–5.
32. Trinity College Dublin (TCD), Price Papers, MS 7538/223, Draft Letter, Dorothy Price to Johannes Heimbeck, 9 January 1948.
33. Anon., 'Hospital Infections', 250; Anon., 'Tuberculosis among Student Nurses', 52.
34. Daniels, Ridehalgh and Springett, *Prophit Survey*.
35. Anon., 'Tuberculosis Infection of Nurses', *BMJ*, 2:3962 (1936), 1212. A memorandum by the Joint Tuberculosis Council stated that radiography was the 'most important method of detecting incipient tuberculosis'. Annual skiagrams rather than tuberculin testing was recommended for nurses.
36. J.B. Lyons, 'Tuberculin Sensitivity: A Review', *Journal of the Irish Medical Association (JIMA)*, 29:171 (1951), 67. From the early 1960s, J.B. Lyons combined clinical work with the writing of medical history. In 1975, he was appointed Professor of the History of Medicine in the Royal College of Surgeons in Ireland. Peter Froggatt, 'A Tribute', in Davis Coakley and Mary O'Doherty (eds), *Borderlands: Essays on Literature and Medicine in Honour of J.B. Lyons* (Dublin: Elo Press, 2002), 1–2.
37. Lyons, 'Tuberculin Sensitivity', 67.
38. Daniels, Ridehalgh, and Springett, *Prophit Survey*, 160–1, 182–3.

39. Ibid., 161.
40. Ibid., 213–4.
41. Ibid., *Prophit Survey*, 202.
42. Evelyn C. Pearce, Correspondence, 'Nurses and Tuberculosis', *BMJ*, 1:4401 (1945), 680.
43. Leading Article, 'The Prophit Survey', *BMJ*, 1:4563 (1948), 1,189–90.
44. R.W. Parnell, Correspondence, 'General Use of Tuberculin Test', *BMJ*, 2:4566 (1948), 107.
45. Bryder, *Below the Magic Mountain*, 242; Medico-legal Correspondent, 'Tuberculosis Contracted by Nurse. Appeal Succeeds', *BMJ*, 1:4755 (1952), 443.
46. Dr [?Edward] Mitchell, 'Nurses' Medicine Lecture', *INM*, 12:13 (1942), 3–4; A lecture given by Dr Mitchell to nurses attending a postgraduate course in Dr Steeven's Hospital, Dublin, included the information that 'among civilised races' certain races are more susceptible to TB with the Irish being the most susceptible and the Jewish the least.
47. Dorothy Price 'Tuberculosis in Adolescents', *Irish Journal of Medical Science (IJMS)*, 6:159 (1939), 124–9; Anon., 'Doctor's Plan to Lower T.B. Death Rate, *Irish Press*, 13 April 1939.
48. H.E. Counihan, 'BCG Vaccination of Student Nurses', *IJMS*, 31:2 (1956), 85. Tuberculin testing was commenced in St Laurence's Hospital, Dublin, in 1947 and BCG introduced in 1949. Morbidity among tuberculin negative nurses was dramatically reduced.
49. Anon., 'Casual Attitude to Tuberculosis', *Irish Press*, 1 July 1939.
50. Alec Wingfield, Correspondence, 'Nursing of Tuberculosis', *BMJ*, 2:4382 (1944), 867; Peter W. Edwards and A. Clark Penman, Correspondence, 'Nursing of Tuberculosis', *BMJ*, 1:95 (1945), 3.
51. Peter W. Edwards and A. Clark Penman, 'Primary Tuberculosis Infection in a Sanatorium Staff', *Lancet*, 245:6345 (1945), 429–31.
52. Edwards and Penman, 'Primary Tuberculosis Infection', 431.
53. Ibid.
54. B.M. Dunlevy, 'Pulmonary Tuberculosis', *INM*, 13:63 (1946), 12.
55. Interview, Mrs Bridie G.
56. Ibid.
57. See F.B. Smith, *The Retreat of Tuberculosis 1850–1950* (London, New York and Sydney: Croom Helm, 1988), 194–203; Linda Bryder, 'We shall not find Salvation in Inoculation: BCG Vaccination in Scandinavia, Britain and the U.S.A., 1921–1960', *Social Science and Medicine*, 49 (1999), 1157–67; Georgina D. Feldberg, *Disease and Class: Tuberculosis and the Shaping of Modern North American Society* (New Brunswick, NJ: Rutgers University Press, 1995), 125–52.
58. Cited in leading article, 'BCG Again', *Tubercle*, 30:1 (1949), 1–2.
59. F.R.G. Heaf, 'Present Trends in Tuberculosis', *IJMS*, 6:252 (1946), 763.
60. Anon., 'Deputation to the Ministry of Health', *BJN*, 94:2137 (1946), 94; Anon., 'BCG Vaccination', *BJN*, 97:2169 (1949), 34. In 1948 Norway 'became the first country to have a law in connection with BCG vaccination of all tuberculin negative members of the population'.
61. Smith, *The Retreat of Tuberculosis*, 201.
62. Leading article, 'BCG Again', *Tubercle*, 30:1 (1949), 1–2
63. Anon., 'BCG Vaccination against Tuberculosis', *BJN*, 97:2174 (1949), 88.

64. Lyons, 'Tuberculin Sensitivity', 67.
65. Ibid.
66. Michael P. Flynn and J. Cyril Joyce, 'The Results of a Tuberculin Study in Rural Ireland', *Tubercle*, 35:11 (1954), 270–6.
67. Anon., 'Tuberculosis among Nurses', *Irish Times*, 28 July 1951.
68. *National BCG Committee Report for year ended 31 Dec. 1954* (Dundalk: Dundalgan Press, 1955), 5. In July 1955, the British MP, Dr Barnett Stross, tendered an apology in the House of Commons for his previous misunderstanding of the situation.
69. Anon., 'T.B. is not Increasing', *Irish Press*, 29 September 1954.
70. Anon., 'BCG can Protect Emigrants', *Irish Times*, 27 October 1954.
71. Jones, *'Captain of all these Men of Death'*, 222.
72. *National BCG Committee Report for year ended 31 December 1955* (Dundalk: Dundalgan Press, 1956), 4.
73. *National BCG Committee Report for year ended 31 December 1955*, 5.
74. Interview, Mrs Bridget E., County Mayo, 30 November 2011. Mrs E. trained as a nurse at Rushgreen Hospital, Romford, Essex between 1954 and 1957.
75. Interview, Mrs Brigid E.
76. Lyons, 'Tuberculin Sensitivity', 63.
77. Evelyn V. Hess and Norman MacDonald, 'Pulmonary Tuberculosis in Irish Immigrants and in Londoners. Comparison of Hospital Patients', *Lancet*, 264:6869 (1954), 132–7; Special Correspondent, 'Vaccination of Irish Emigrants', *Irish Times*, 17 July 1954. Hess and MacDonald concluded that susceptibility was due to low exposure with social factors playing a role.
78. Hess and MacDonald, 'Pulmonary Tuberculosis in Irish Immigrants and in Londoners', 137.
79. G.Z. Brett, 'Pulmonary Tuberculosis in Immigrants: A Mass Radiography Study', *Tubercle*, 39:1 (1958), 24–8.
80. V.H. Springett, J.C.S. Adams, T.B. D'Costa and M. Hemming, 'Tuberculosis in Immigrants in Birmingham, 1956–1957', *British Journal of Preventive and Social Medicine*, 12 (1958), 135–40.
81. N.J. England, Chairman Joint Tuberculosis Council, Correspondence, *BMJ*, 1:5076 (1958), 941–2.
82. M. Dunlevy, 'Childhood Tuberculosis in the Mid-Century Decade', *JIMA*, 44:261 (1959), 76–81.
83. Hardy, 'Reframing Disease', 535.
84. Welshman, 'Importation, Deprivation', 142–3.
85. Health Protection Surveillance Centre, *Tuberculosis Guidelines 2010*, 114.
86. Power, Sim, Gallagher and Greiner, 'A Study to Compare Chest X-ray Reports on Overseas Nursing Recruits', 140–1.
87. Welshman, 'Compulsion, Localism', 295–312. Welshman discusses the possibilities of screening immigrants for TB before departure, on entry and after arrival in the United Kingdom between 1950 and 1965.

6
Immigration, Ethnicity and 'Public' Health Policy in Postcolonial Britain

Roberta Bivins

What happens at the end of empire, when decolonisation draws far-flung populations 'home' to an environment – political, social and physical – very different from either colonial or imperial expectations? Drawing on British examples, this chapter explores medical and especially public health responses to postcolonial migrants and their ethnically marked descendants in the era of decolonisation and Cold War, responses often generated by professional men and women themselves returning from or building upon careers begun in Britain's tropical colonies. It focuses on two diseases which came to be closely associated with immigration in the years after the Second World War: tuberculosis (TB) and rickets (in adults, osteomalacia). This pairing facilitates comparisons between medical policies and projects mediated largely by public health actors and interventions, and those shaped principally by the interests and innovations of elite biomedical research. Additionally, it presents the different ways in which infectious and nutritional disorders were addressed and the impact of 'race' on perceptions of 'imported' illnesses and the migrants affected by them.

Both rickets and TB were familiar to medical authorities in Britain and both were readily curable in the post-war period. TB was, by the 1950s, a well-understood bacteriological condition sharply declining in prevalence. Nonetheless, with medical and public hopes pinned optimistically on its eradication, TB remained the focus of considerable public health activity within the majority community (for the purposes of this chapter, defined as white, largely Christian British-born descendants of white, largely Christian British-born parents) as well as migrant minority populations.[1] Its problematic association with migrant groups – first the 'susceptible' Irish, then the 'suspect' South Asians – and the distinctive ways in which the latter group was regarded in comparison with their predecessors and the majority population reveal professional and political tensions within British medicine and government. Comparing reactions to majority, Irish and South Asian at-risk populations also highlights the ways in which 'race' and 'ethnicity' inflected responses to the twin markers of nativity and health status.

In contrast to TB, 'normal' (that is, vitamin D deficiency) rickets had essentially disappeared from the majority community during the Second World War, not least because its treatment was well understood, inexpensive and easy. The wartime regulation of the food supply, mandatory fortification of basic foodstuffs and provision of free supplements for vulnerable populations reduced the nutritional inequalities associated principally with poverty and social deprivation, and specifically targeted dietary deficiency diseases including rickets. This played a fundamental role in rickets' near-eradication, which in turn became a source of pride not only for those medical professionals and researchers who actively shaped British food policy during the Second World War, but for the general public who saw the disappearance of this 'disease of poverty' as a sign of triumphant modernity.[2] However, the causes of rickets remained ill-defined, and only became better understood when elite researchers came into contact with a new and apparently compliant pool of 'clinical material': immigrant and second-generation sufferers of what was quickly termed 'Asian rickets'. As the medical community debated the relative importance of diet and behaviour, environment and skin pigmentation in causing migrant susceptibility, they recapitulated colonial medical debates about the respective roles of culture and race in 'native' pathology. In much the same way, medical discussions of migrant TB reinvigorated its image as a 'disease of civilisation', simultaneously reinforcing the well-established image of the infectious immigrant. Medical considerations of both diseases spliced older models of environmental risk and fragile racialised bodies with newer understandings of assimilation as a medical prophylaxis. Thus close study of medical responses to ethnic minority communities in post-imperial Britain offers considerable traction on historical understandings of and approaches to the 'postcolonial'. It was well-recognised that the new migrants were poor as well as 'ethnic'; however, poverty itself was an insignificant part of either TB or rickets discourse in post-war Britain, for reasons explored in more detail in John Welshman's chapter in this volume, and also because officials and many medical professionals saw the availability on the National Health Service (NHS) of free medical screening and treatment (for TB) and free or low-cost supplements (for rickets) as eliminating potential *economic* barriers to health. Only in relation to poor housing stock did health officials and doctors explicitly cite poverty as a factor in poor migrant health; even here, the principal concern was with the transmission of infection among migrants due to overcrowded conditions, rather than the grinding impact of impoverishment on physical and mental health more generally.

Scrutinising responses to specific marginalised groups also offers traction on the nature of public health in Britain in the post-war, post-NHS period. This chapter examines conceptions of 'the public' as defined by public health practitioners, policy makers and politicians. Were Britain's immigrants and their children, particularly those who were both racialised and ethnically marked, fully incorporated by such medical and political conceptions? Did policy makers and medical professionals consider migrants and their descendants as an integral part of the populations they both scrutinised and served or were these groups erased, submerged or marginalised? This chapter will test claims that the limited assimilation of migrants of Asian origin or descent was seen as beneficial to public health. Finally, the pairing of an infectious and, in immigrants, often acute condition with a chronic non-infectious nutritional disorder should cast some light on the changing nature of 'public health' in the post-antibiotic era, as increasing clinical attention focused on individual idiosyncrasy, whether genetic, biochemical or 'cultural'. Could a model of 'public health' incorporating large-scale centrally directed interventions survive this increasingly individualistic vision of 'the public'?

'Let's stamp it out': Tuberculosis, eradicationism and the immigrant 'Threat' after the Second World War

As Anne Hardy has argued, the exigencies of military mobilisation prompted new attention to the problems of TB in the British population during the Second World War. Mass miniature radiography (MMR) and tuberculin testing emerged as crucial tools of population surveillance.[3] They were at the heart of anti-TB campaigns and control strategies. At the same time, better nutrition and housing, new social support systems, including allowances and rehabilitation services for those affected and compliant with treatment, and a rapid increase in hospital beds for TB patients played key roles in pushing TB infection to the margins of British society, where it survived largely among the very young and the very old.[4] School health services and feeding programmes, and (from

1953) BCG vaccination targeted the young. Moreover, in the immediate post-war period, Britain's veterinarians effectively eliminated TB from the national herd, removing a major indigenous source of environmental risk. Of course, most strikingly, post-war experiences of TB were transformed by the advent of effective antibiotic chemotherapy. Taken together, these factors contributed to the widespread popular and professional conviction that TB, like smallpox, would disappear from Britain within a few years. In this climate, MMR directors and chest specialists, likened by Ministry of Health civil servants to the biblical Gadarene swine rushing towards the cliff edge of eradication, faced increasing pressure to bring Britain's remaining loci of infection under control.[5]

In 1957 the Scottish Department of Health initiated a massive campaign to clean up one of the nation's most notorious TB 'blackspots': Glasgow. The city's rates of TB were the highest in the United Kingdom, at 25 cases per 1,000 residents, and unlike England, Glasgow's TB notification rates were rising, rather than falling. The residents of its densely packed slums – the city was among the poorest in Britain, and some 40 per cent of its population still lived in tenements, often with outside toilets – in particular, were seen as resistant to public health messages. To address this crisis, the Glasgow campaign brought in 37 MMR units, borrowed from health authorities across England and Scotland, and stoked intense media coverage through advanced publicity and a series of attention-getting measures, including fireworks and a weekly raffle offering all screened individuals the chance to win extravagant prizes. Importantly, volunteers raised awareness both of TB and of the impending campaign by house-to-house visits in crucial neighbourhoods. During the campaign's five weeks of intensive scrutiny, 714,915 Glaswegians (76 per cent of the city's total population) voluntarily attended for x–ray and 2,755 new cases of active TB were identified (alongside 5,379 cases where activity was uncertain).[6] The campaign was particularly successful in reaching older men, well-known for their relative unresponsiveness to public health measures.[7] Indeed, on the campaign's final day, a queue of last-minute stragglers spilled out into the city's George Square. A photograph of these latecomers shows a crowd of middle-aged and older men in overcoats.[8] Its wide reach and success in persuading so many to come forward was recognised at the time as the campaign's most salient feature. As the *Lancet*'s editors argued, 'most interest will centre on how such a high proportion of the population were persuaded to come forward'. Within the medical community, a consensus emerged that its 'secret of success' was 'the intensity, ubiquity, and effectiveness of the publicity', which ranged from the posters that hung on buses, tramcars and in three out of every four Glaswegian shops to cinema notices, broadcast and print news coverage and regular prize distributions to participants.[9]

This was a campaign intended to reach an entire population, rather than targeting particular groups perceived as especially vulnerable or threatening. In addition to its use of the media, the Glasgow campaign mobilised local volunteers to motivate their neighbourhoods and to canvass the homes of non-participants. Organisers argued that this two-pronged approach was essential to the city's high rates of turn-out. Contemporaries agreed that the public's enthusiastic response in turn produced a significant reduction in rates of disease.[10] Consequently, Glasgow's success became a model for other British cities.[11] Certainly, the campaign was not cheap; the Scottish Department of Health reported that it had cost £114,269, or 3s 2d per examination (and £16 4s per new TB case identified; in terms of purchasing power, roughly £305 in 2010). Publicity, however, accounted for only £17,000 of this budget, or 5 ¾d per x-ray, 'an expenditure which', as the *Lancet* noted, 'was obviously well-justified'.[12]

As Scotland prepared to 'stamp out' TB in its remaining urban strongholds, experts in England increasingly drew attention to a new source of the disease, perhaps explaining a phenomenon which had begun to trouble the profession: plummeting rates of TB mortality had not been matched by similar declines in morbidity. Specialists had reported high rates of the disease among the Irish migrants drawn to Britain by its hungry labour market in the early 1950s; in the mid-1950s, they speculated that West Indians too would be affected; and by the late 1950s and early 60s, medical authorities regularly expressed alarm at rates of infection among newly arrived South Asians. Capitalising on wider anti-immigrant feeling, some in the tabloid press declared TB, 'the disease people thought was beaten', a continuing threat to the public health.[13] The British Medical Association (BMA, often described as the doctors' union) joined battle with increasingly strident calls for port or pre-departure medical screening and compulsory treatment for migrants. For those already established in the United Kingdom, they demanded improved detection. Yet despite the success of the Glasgwegian and subsequent city-wide campaigns, when high rates of TB were identified among Britain's postcolonial immigrant populations, the introduction of similarly large-scale, resource-intensive appeals was dismissed.

In part, reluctance to deploy mass campaigns to address TB among specific immigrant groups reflected wider changes in central and local attitudes towards such interventions, particularly in relation to a disease fast disappearing from the majority population. J.E. Geddes, Chief Supervising Tuberculosis Physician in Glasgow's busy Tuberculosis and Chest Service, summarised the changes which militated for a different approach: the 'individual and communal significance of tuberculosis' had declined; eradication was now seen as an 'attainable', even inevitable, objective. It was, however, vulnerable to 'easy optimism'. In light of this, Geddes urged the need for both 'ceaseless endeavour' and 'constant awareness' of the changing clinical and epidemiological picture. To succeed, eradication efforts must be 'relevant and purposeful'. Mass screening was revealed by these metrics to be 'outdated'; instead he suggested targeted screening of specific groups: those put at risk by their working conditions; those falling into age and sex categories marked by high prevalence; those 'whose social habits create a hazard' and 'recent immigrants'.[14] The *Lancet* too asserted in 1958 that declining rates of disease were rapidly rendering mass radiography 'unprofitable'.[15] But not all public health workers were so ready to relegate mass campaigns to history. Reporting on Edinburgh's 1958 campaign, for example, the city's MOH and two colleagues declared, 'a campaign of this sort is a useful and justifiable method of case-finding for tuberculosis and other serious chest diseases. It is particularly suitable in communities where the disease appears to be coming satisfactorily under control'.[16]

At the Ministry of Health in London, the cost of mass screening, and particularly the cost per new case identified, seems to have been the principal argument against its use. The alternative was, as Geddes implied, screening targeted at specific groups, including migrants and other groups deemed either to be at risk or to impose risks on others. However, applying selective screening to these populations presented problems as well. Irish migrants had long been analogised to England's disease-ridden, unhygienic and medically recalcitrant underclasses, and this perception undoubtedly persisted in the post-war period.[17] In 1955, medical civil servant Dr Daniel Thomson was matter-of-fact in characterising them as 'of comparatively low intelligence and particularly wayward in habit'. He asked rhetorically: 'is it not that the Irish immigrant (maybe to an exaggerated degree) represents that stratum of society in this country to which the impetus of our anti-tuberculosis drive – by local authorities and by radiological units – must increasingly be directed?' At the same time, Thomson's casually derogatory attitude also reveals the extent to which the Irish, at least, were incorporated into the 'public' of British public health. Invisible the Irish might indeed become to Britain's medical gaze, but even when suspected of importing TB (as they originally were in this period) the Irish were a known quantity and an established constituency – albeit an irresponsible, unheeding and wilful one – of British public health measures.[18] Thomson concluded that, for this group, the solution was a simple one: 'In coping with the general, the particular problem will be solved'.[19] Special measures to address TB among Irish immigrants were unnecessary, because their needs and the risks they imposed would be captured in the existing net of medical services and surveillance. This basic assumption would inform the Ministry's first-line responses to migrant health problems until the 1980s.

Yet by the end of the decade, the mass migration of workers from Britain's tropical colonies and the new nations that emerged in the wake of the receding of empire changed the tenor of debate (if not the Ministry of Health's preferred response). The picture of migration into Britain in the immediate post-war period is complex. It incorporates a peak of 'alien' migration from Europe – largely as European Voluntary Workers (EVWs) – in the late 1940s and early 1950s; a spike of Hungarian refugees in the mid-decade; and a high and relatively consistent level of Irish migration, dominated by young men entering the building and construction trades, and young women working in the expanding service sector or nursing posts. The Irish population was fluid, with individuals often migrating to take up seasonal work and returning to Ireland in the off-season. From the 1948 docking of the ship the Empire Windrush at Tilbury to the end of the 1950s a steeply rising number of migrants from Britain's Caribbean colonies attracted much attention. While this 'flood' of non-white colonials stirred moral panic, West Indians contributed a relatively small proportion to overall immigration which was dominated by the Irish (by 1961, the census enumerated 900,000 Irish-born UK residents; current research suggests that it underestimated their numbers by as much as 20 per cent).[20] Finally, from roughly 1956, South Asians, and especially Pakistanis, emerged as a rapidly growing migrant group – again, triggering a reaction disproportionate to their actual numbers.[21] Of these populations, only the 'alien' Europeans were initially subject to immigration controls. The first legislation controlling the entry of Commonwealth citizens and colonials to Britain passed into law in 1962, as the Commonwealth Immigrants Act.

Throughout the decade, Irish migrants accounted for by far the greatest proportion of non-indigenous cases of TB; in 1955, for example, there were more Irish migrants undergoing treatment for TB in London's North West Hospital Region than all other migrant groups combined.[22] However, as evidence emerged suggesting that a majority of Irish migrants contracted their TB in the United Kingdom, the focus of professional and media attention shifted to an even more contentious group.[23] As John Welshman has documented, migrants of West Indian and South Asian origin were swiftly blamed for expanding Britain's 'infector pool' and threatening eradication efforts.[24] Despite strong resistance from many officials and medical experts in the Ministry, the concatenation of anti-immigrant sentiment and racial bias not only promoted efforts (ultimately unsuccessful) to medicalise and restrict immigration but drew intense criticism of the Department of Health's 'laissez faire' response to immigrant health.[25] While the Department continued to argue, throughout the 1950s, that 'imported illness' was a minor problem and required neither new legislative controls on migration nor additional public health measures, they were soon forced to address wider public disgruntlement about the absence of meaningful health controls (and what ministers described as 'undue publicity to the problem').[26] In 1955, Patricia Hornsby-Smith, then a junior Minister for Health, was forced to rebut Parliamentary critics led by Dr Barnett Stross. At the time, she presented the Ministry's inaction on TB among migrants as arising from reluctance to give any single group 'special treatment'. Instead, she offered evidence on the incidence and control of TB among migrants, suggesting that they posed no danger of 'seriously jeopardising our record in development against tuberculosis' or undermining 'the progress we have made'.[27]

In reiterating Britain's continued success in reducing TB incidence, Hornsby-Smith was explicitly addressing public anxiety that immigration impeded progress. The admixture of what were perceived to be pre-modern peoples with Britain's modern indigenous population might, some suspected, endanger Britain's modernity as much as its health. These concerns were reflected in frequent public complaints about the migrants' 'primitive' practices and poor hygiene, and their supposed negative impact on public spaces and services.[28] Nevertheless, the Ministry's politicians repeatedly dismissed the idea of a 'cordon sanitaire' as impractical and unsuitable. Establishing another pattern that would become a familiar part of the Ministry's response to complaints about immigrant health, Hornsby-Smith emphasised that these were local problems in need of local attention and solutions. They were not matters for top-down central governmental action, which she hinted would only make matters worse: 'The best method is that boards and local authorities which have this concentration of immigrants and are alive to the conditions in their area should deploy its mass radiography units so as to seek out the danger spots'.[29]

The idea that immigrants represented a health threat only in highly localised sites reflects an emerging understanding that 'coloured' or 'colonial' migrants in particular formed discrete 'colonies', isolated from the British majority. Indeed, some argued that the perceived failure of migrants from India and especially Pakistan to integrate was positively beneficial to the public health. As one doctor put it in 1964, 'a greater degree of integration with the indigenous community would increase the risk of dissemination'.[30] In other words, the enclavism which had sheltered Europeans from 'native' disease in South Asian and Africa was still protective when the polarities of migration (but not of power) were reversed, and racial discrimination in UK public and private housing forced the new migrants into dangerously overcrowded enclaves of their own.[31] At the same time, the scanty evidence of one failed mass radiography campaign directed specifically at these migrants in the Midlands manufacturing city of Birmingham was sufficient to convince the central planning team that they also would not respond well to targeted measures.[32] Instead, the Ministry urged – but did not itself initiate – a new protocol of disguising targeted local measures as general ones: 'The local authority should organise the general publicity and preventive medicine campaign so as not to single out this particular group and make it feel that it is being attacked as containing tuberculosis suspects, and so scare them off'.[33] The perceived need for such dissembling highlights the degree to which this migrant population was not considered as a constitutive element of 'the public'.

Ultimately, while the BMA's campaign in support of strict, exclusionary medical controls either in migrants' countries of origin or at the United Kingdom's own borders failed to achieve those primary goals, South Asian migrants in particular came to be identified closely with TB, and particularly with 'imported' TB. By 1965, the idea that these particular in-comers posed a threat both to British bodies and to British modernity was well-known and had a profound impact not only on the host community but on the migrants' perceptions of themselves. In 1965, the *Listener* magazine – published by the BBC and intended as the literary and intellectual complement to its more populist listings magazine, the *Radio Times* – printed an article on the new immigrants that included quotations from broadcast vox populi interviews. The comments of one interviewee reveal the extent to which this discourse of infected immigrants permeated wider discussions of immigration and diversity as a whole. Speaking of her life in Britain, a 'beautifully garbed Indian girl' was quoted as saying: 'The difference here...is that prejudice is rooted in reason. When people say..."They're dirtying our streets", "They're bringing T.B. into the country", it's because they are. That's not prejudice. It's xenophobia'.[34] Her words emphasise the outsider status assigned to Asian immigrants, her own awareness of that status, and an apparent acceptance that the prejudice she and others in her community faced – at least to the extent that it arose from fears of contagion – was 'reasonable'. But what about British popular, professional and political responses to non-infectious diseases? When postcolonial, racialised migrants presented no 'threat' to the health of their hosts, how were their medical concerns conceived and addressed in post-war Britain?

'Asian rickets', imported opportunities and the emerging 'Multicultural' state

Like TB, rickets was historically a public health concern in Britain. Long known as 'the English disease', the form of rickets caused by simple deficiency of vitamin D was virtually eradicated in Britain during the Second World War. Amid deep concerns about population nutrition during the war, Whitehall ministries operated a three-pronged strategy of intervention. First, the state controlled the national diet via rationing access to foodstuffs, and the provision of special diets and supplements to particular groups (for example, expectant and nursing mothers, infants, young children and hospital patients). Simultaneously, manufacturers were required to fortify flour with calcium, margarine with vitamins D and A, and to mill only bulky, nutrient-rich high-extraction flour. Alongside this regulatory regime, the state avidly sponsored education in nutrition, weaving messages about health and diet throughout its rich propaganda programme.[35] These policies were strongly supported by the Medical Research Council under biochemist Edward Mellanby. Post war, there was some continuation of these policies via the provision, through the NHS and maternal and child welfare feeding programmes, of free or heavily subsidised supplemental foods (e.g. cod liver oil) to groups deemed at risk.[36] However, mandatory fortification in the manufacture of foods was retained only in relation to infant formulae and margarine. With the incidence of rickets and osteomalacia vastly reduced and all basic foods available and relatively affordable to an increasingly affluent society, emphasis turned instead to health education. Direct state or public health interventions into individual nutrition were deemed appropriate (in the new era of free universal health care) only at the level of the clinical encounter, and via the provision of an ever-shrinking nutritional safety net (Welfare Foods) accessible to the poorest families. Unlike the United States, in Britain, popular enthusiasm and market forces did not drive near-universal commercial fortification of basic foodstuffs like fluid milk or breakfast cereals. Instead, vocal (minority) opposition to 'mass medication', whether via vaccination or fluoridation, alongside the United Kingdom's well-established food purity legislation, justified regulatory withdrawal from the nation's kitchens and offered little incentive for manufacturers to fill the vacuum voluntarily.[37]

The disappearance of deficiency rickets unmasked previously unrecognised metabolic and genetic disorders that manifested clinically as 'resistant rickets'. Biochemistry and genetics were internationally recognised as leading-edge research areas in the biomedical sciences; in the late 1950s and 1960s, their high status made research on these metabolic nutritional disorders attractive to ambitious researchers in Britain – and to Britain's no less ambitious Medical Research Council.[38] However, the conditions themselves were rare; moreover, although the various forms of metabolic rickets allowed researchers to probe the complex biochemistry of nutrition in new ways, their investigations could only offer speculative evidence about the functions and absorption of micronutrients in the 'normal' body. As elite researchers tested both hypotheses and therapies on the wards, nagging doubts remained: could their findings be generalised to the entire human population, or did they apply only to the unusual and idiosyncratic individuals who populated their specialist units and wards? Charles Dent of London's University College Hospital Metabolic Ward reflected on this problem in 1956, worrying about a new synthetic treatment, DHT (dihydrotachysterol) that apparently treated metabolic deficiency successfully but reputedly could not heal 'true rickets'. He urgently sought opportunities to try the compound in a nutritional rickets case. Unfortunately, Dent found it nearly impossible to identify any such cases among the majority population. His proposed trial stagnated until a migrant child showing symptoms of rickets was referred to his unit in 1960:

> The family being Turkish Cypriots makes me think very strongly that we may have an instance here of classical rickets due to oral dietary deficiency, undoubtedly the rarest cause of rickets nowadays in British people...curiously enough we see it much more often in visiting imigrants [sic] from backward countries.[39]

The migrant groups most severely affected by and closely identified with rickets were those of South Asian origin. As Dent and other researchers rapidly discovered, these growing populations offered British biochemists a reservoir of archaic conditions and a laboratory for public and individual interventions.[40] Key research groups among Britain's elite biochemists, based in London, Manchester and Glasgow, engaged intensively with their local communities, creating in the process the new disease entity 'Asian rickets'. For some, this disease represented simply a funding opportunity and a new avenue by which to approach a knotty problem at the cutting edge of biochemical research. S.W. Stanbury, head of the University of Manchester's new metabolic medicine unit, was clear about his priorities, informing the Department of Health that his team 'would be reluctant to undertake further population work unless the effort returned information relevant to our main themes of research ... we are primarily a biochemically orientated clinical research team'.[41] For others, 'Asian rickets' became very nearly a crusade to change public health responses to nutrition, or to better integrate an isolated and under-served population into British ways of life.

For the Department of Health, meanwhile, 'Asian rickets' was a minefield. The return of rickets provoked controversy in part because it was an affront to public expectations and eroded political and medical achievements. Thus in 1967, W.T.C. Berry, then Secretary of the Committee on Medical Aspects of Food Policy (COMA, the Ministry's chief fact-finding and advisory panel on nutrition and fortification) and a former colonial medical officer familiar with the far more evident malnutrition of West Africa, testily declared to the Nutrition Society's Annual Meeting, 'Rickets excited great emotion, largely because it is thought that it is reappearing and this is a sign of social regression'.[42] Berry clearly regarded this 'emotion' as excessive, dismissing the 'attention' given to 'the minutest signs of possible vitamin D deficiency'.[43] Berry was not the only sceptic. Clinicians and researchers, as well as politicians and the press, expressed profound uncertainty about the nature, incidence and aetiology of the condition; some patients certainly displayed frank rickets, with bowed legs, the 'rickety rosary' and deformed pelves of severe deficiency. But for many migrant children and adolescents, the disease was defined biochemically, rather than clinically or radiologically – and not all authorities accepted that this more subtle deviation from 'normal' health represented a condition in need of surveillance or treatment.[44]

At the beginning of the 1970s, the return of rickets was also a hot potato because of the ease with which it could be imbued with party-political meaning. However, with ideas of 'race relations' still in their earliest stages, it was initially the possibility that rickets might be recurring among the poorest sectors of the majority population that gave the condition significance. When the Tories returned to power in 1970, spending cuts were an immediate priority; the frayed remains of the Welfare Foods safety net seemed an easy target. In 1971, Margaret Thatcher, as Education Minister, cut the provision of school milk to children older than seven. Although unfortified liquid milk was not a significant source of vitamin D, this point was lost in the ensuing uproar; opposition parliamentarians immediately blamed rickets' apparent resurgence on the 'milk snatching' Government. Ernest Fernyhough, Member of Parliament (MP) for Jarrow, complained: 'It was the proud boast of this country in the middle 1950s that, because of our welfare and health services, British doctors had to go abroad to study rickets because none of our children were suffering from it'. He accused the government of pursuing policies that would make rickets again 'commonplace in the poorer districts'.[45] A Labour Whip, meanwhile, used rickets to draw attention to the Conservatives' history of reducing the provision of nutritional supports:

> the incidence of rickets had increased after the reduction of the Vitamin D content of welfare foods and the increased cost of National Dried Milk between 1956 and 1960...Rickets was a disease that had disappeared, but it is now reappearing...That is some indication of the long-term effects of the sort of policies that are crystallised in the [Education (Milk)] Bill.[46]

By 1973, research in cities across the United Kingdom traced a clear rise in the incidence of rickets – but offered scant and highly contested evidence of the condition among Britain's majority population. A heated Parliamentary exchange between Thatcher and Labour MP Laurie Pavitt strikingly demonstrates the key role of politics in bringing an easily cured and still uncommon non-infectious condition into the media and political spotlight. Pavitt accused Thatcher head-on: 'there is now virtually an epidemic of a new form of rickets, biochemical rickets...Is it not time that she reversed her policy on school milk?' Thatcher's rebuttal reveals the degree to which immigrants remained unthinkingly excluded from the 'public' of public health. Admitting that she had seen the results of a Birmingham study that demonstrated a clear rise in incidence even of radiological rickets, she nonetheless dismissed it: 'All 24 [affected pupils] were immigrant children'.[47] Thatcher's cuts remained in place.

As well as the cut and thrust of party-politics, responses to questions of immigrant (and, with the growth of British-born children of Asian descent, ethnic minority) health were shaped by wider trends in social and medical thought. By the 1970s, both the grounding principles and the structures of public health were changing. Post-war shifts in balance between social medicine and 'risk' models of public health stimulated the development of new interpretations of 'national' and 'public' health. Moreover, the NHS and the Ministry of Health (amalgamated with the Ministry of Pensions in 1968 to become the cumbersome Department of Health and Social Services (DHSS)) lost ground financially relative to other spending ministries and DHSS sub-departments. In line with these trends, central health authorities resisted ever more fiercely pressure to instigate national campaigns or to take nationwide action on matters of public health. Instead they urged the greater suitability and sustainability of local efforts. In the case of 'Asian rickets', this approach produced a highly variable response, ranging from intense efforts at targeted nutrition education in Lancashire to no action whatsoever in some of the hard-pressed London boroughs. The dangerous effects on ethnic minority groups (and indeed minoritised groups of all descriptions) of this broad central delegation of responsibility and costs to 'local provision' would rapidly become evident. Nonetheless, writing in 1977, as the Department faced growing criticism of its inaction on ethnic minority health, nutrition officer Sylvia Darke set out the principles which underpinned its inertia:

> *Alterations in national policy are in general reserved for problems which affect the national health* and which can only be solved by Government action, for example, a rationing system or welfare policy ... It is as well to remind ourselves that the public health means the health of *56,000,000 people.*[48]

The health problems of minority populations were not, by this standard, suitable for national intervention. Moreover, monitoring nutrition status was both challenging and expensive, in part because biomedicine and public health increasingly focused on the individual, at least in the 'developed' world: '[w]hen ... a sufficient food supply is assured there is no simple yardstick by which to judge nutritional status from dietary intakes alone and *a medical assessment of each individual becomes necessary'.*[49] When the only marker of insufficiency was a non-notifiable condition, arguing for such costly and minute investigations was difficult. Indeed, internal memos show that Darke herself pleaded within the Department for more funding to campaign against 'Asian rickets', but was rebuffed.[50] The evidence available 'confirmed that the problem of rickets is confined chiefly to Asians in urban areas': central action was not merited, and the broadcast campaigns Darke advocated were dismissed by her superiors as too expensive for such a 'very small target'.[51] Yet local and regional authorities faced financial retrenchment, the fragmentation of local public health services and the replacement of often locally and regionally influential Medical Officers of Health with posts in the still-nascent specialty of 'Community Medicine'.[52] Almost the only central action approved and initiated was the preparation of a pamphlet to alert medical professionals to the risk of rickets among their patients of South Asian descent.

This document, drafted in 1975–76, is revealing. It placed a heavy emphasis on the absence of vitamin D fortified foods or vitamin supplements in 'Asian' diets, but also noted a peculiar Asian 'susceptibility' to rickets 'whether for genetic reasons, or because of their diet and racial customs'. Skin pigmentation itself was ruled out as a cause by the rarity of rickets among West Indians living in Britain.[53] A significant theme was 'the special problems of immigrant families'. Through comparisons with other migrant groups, the authors drew one clear conclusion: assimilative behaviours were medically protective, at least in the case of rickets, and the perpetuation of traditional practices, particularly of diet and dress, was unhealthy.[54] Thus in this document, aimed at medical professionals ranging from clinicians and public health workers to health visitors and GPs, Asian women in particular shouldered the blame for their children's rickets, and the pathogenic refusal to assimilate, perpetuating colonial gendering of failed modernity. It made no critical comment about the failure to provide useable services for this 'vulnerable' group. Instead, it defended the British climate, through re-positioning rickets and osteomalacia as 'tropical' conditions caused by the practices of purdah and seclusion. In other words, 'Asian rickets' sufferers were not innocent victims of the gloomy climate or governmental neglect, but were responsible for their own illnesses to the extent that they failed to assimilate.[55]

Over the course of the 1970s, however, central inattention to ethnic minority health attracted both media anger and critical attention from Asian communities themselves. The new Community Relations Commission joined forces with the British Nutrition Foundation to offer its own critique.[56] Medical professionals – often experts 'colonised' by the patients who had become so vital to their research and public health professionals – also clamoured for a change in approach.[57] In 1977, even the BMA's Annual Representative Meeting called for prophylactic fortification. Moreover, activists drew explicit attention to the difference between the state's overwhelming efforts to address rickets in the 1940s, when it had been a disease of the majority community, and the 'ad hoc nutritional advice' it provided to ethnic minority sufferers.[58] Inaction on rickets was becoming inaction on 'race relations'. In short, the political climate was changing, and politicians in the United Kingdom, like their American counterparts at the beginning of the decade, felt pressure to demonstrate interest in ethnic minority health concerns.[59] As an internal memo put it in 1980, 'expectations had been aroused, attention had been drawn to the difficulties faced by health authorities in providing effective health education across language and cultural barriers, and the Department's non-involvement was increasingly criticised'.[60] 'Asians' (no longer necessarily 'immigrants') were gradually gaining a place in 'public health'.

This was the context in which Gerard Vaughan, the new Minister of State for Health (and himself medically qualified), was inspired to push for a public campaign actively involving Asian communities and 'leaders'. Vaughan's 'Stop Rickets' campaign was designed to showcase engagement with long-ignored communities, and simultaneously to demonstrate the cost-effective power of education-only campaigning as compared to 'nanny-state' regulation and 'mass medication'. Vaughan's arrival at the DHSS had coincided with a period of prolonged media pressure for action on rickets (among other health inequalities), and almost immediately on arrival, this pressure began to tell. At the end of July 1979, Vaughan met with key Departmental staff working on nutrition, determined to tackle the torpid rickets policy. His urgency was a direct response to what he termed the 'disaster' of a Granada TV's 'World in Action' programme on the subject, which promoted fortification and presented the department as 'doing little'. He called for a meeting with 'the leaders of the Asian community' and their co-option to 'stimulate a greater awareness amongst their communities' of already-available services to prevent rickets and promote healthy nutrition.[61]

But Vaughan's campaign was more than just ideological showboating. Significant progress had been made in the decades since rickets' rediscovery, both in terms of understanding the disease itself and in terms of attitudes towards the communities most affected. While the emphasis on identifying 'community leaders' through whom to act – rather than attempting to engage directly with community members, and particularly the women who were most directly affected both by the disease and by the changes required to eliminate it – resonated

uncomfortably with colonial practices of indirect rule, this was nonetheless a striking improvement on earlier efforts, in which assumptions
about cultural preferences and barriers substituted for knowledge and
consultation.

'Stop Rickets', for the first time, also demonstrated a growing realisa-
tion that medical professionals and civil servants needed partners on the
ground to effectively address problems of ethnic health. Despite considerable opposition from his own civil servants, Vaughan installed a
Working Group that included a significant number of representatives
from national British Asian organisations. These new partners were,
perhaps unexpectedly, influential. Nonetheless, the resulting 1981–82
campaign was very much a compromise. The representation of key
ethnic minority groups in the planning stages was a significant innovation at the national level (some local authorities had pioneered this
approach) and forced the DHSS to invest more heavily in the effort
than it had initially planned. However, 'Stop Rickets' remained at the
national level an 'education-only' effort. Within the Department, there
was active distrust of the Working Group's Asian members; even the
advertisement for a campaign director listed knowledge of the NHS as
'essential', while knowledge of 'Asian cultures and an Asian language'
was considered only 'highly desirable'. Yet in the end, Veena Bahl was
appointed instead of an internally preferred NHS insider, sending a
strong visual and political message; the fact that her post was to be situated within a charity, Save the Children, rather than the DHSS itself,
was likewise intended to signal a new institutional approach.[62]

Ironically, Vaughan's campaigns also demonstrated the persistence of
top-down centralised modes of planning and intervention so typical
of colonial and postcolonial medicine in the former colonies themselves. When local areas suggested either that the rickets campaigns
were unnecessary (because rickets and osteomalacia were only comparatively minor elements in the bleak ethnic health picture) or duplicative
of existing remedial efforts, they were largely ignored. Instead, the DHSS
and 'Stop Rickets' campaign plunged ahead. When its materials were critiqued as undermining local – and often far more nuanced – initiatives,
efforts were made to placate local interests rather than to adapt to local
needs. Tensions between different branches of the medical professions
likewise came into the fore, particularly through rivalries between doctors and nutritionists, both of whom felt that they were the experts in
matters of diet and nutrition education.

So did the education-only approach succeed and for whom? Evidence suggests an ambivalent picture. On one hand, the 'Stop Rickets'

Figure 6.1 'Healthy Bones', Lancashire Area Health Authority, c.1980

Produced by the Lancashire Area Health Authority in the early 1980s, this Urdu leaflet was tailored to the wider nutritional needs and dietary practices of the local British Pakistani population. Forwarding examples of her educational materials on to the DHSS in 1981, the district dietician wrote in exasperation: 'I could tear my hair out when I think of all the frustration...I went through trying to get the money and support for our work, and now they are going to spend thousands just on rickets!' The National Archive MH160-1453 Letter, Margaret Allen (District Dietitian, Lancashire Health Authority) to Pat Torrens (DHSS Advisor on Dietetics), 9 February 1981. This image is licensed under the terms of the Open Government Licence.

campaign raised and improved the profile of the DHSS and its services among British Asian communities; likewise, it exposed medical professionals and civil servants, as well as the public, to the deprivation faced by these diverse populations and to the resources that such communities were nonetheless able to bring to bear. Surveys performed to assess the impact of the campaigns were able to show a positive effect on knowledge about rickets and its prevention in the immediate aftermath of the campaigns. The *British Medical Journal* (*BMJ*) proclaimed it a success, both in terms of its public health impacts and for its 'beneficial side effects' of improving communication between the DHSS and ethnic communities and – more tellingly – 'race relations' in general.[63] It also gleefully reported the astonishingly low levels of resources committed by central government to solving what they themselves presented as a public health emergency: one local campaign cost the Department only £250 (approximately £750 in 2012), although campaigns in the largest communities cost more.[64] On the other hand, a greater proportion of departmental resources were deployed on producing educational materials, reflecting a renewed focus on reducing barriers to accessing information about health and indeed to the NHS and allied social services. Nonetheless, a year later, the difference between communities which hosted and which had not experienced the campaigns were marginal. Moreover, of course, knowledge about health does not necessarily correlate to improved health status.[65]

Did 'Stop Rickets' change professional or political perceptions either of rickets or of Britain's Asians? The public-facing rhetoric of Vaughan's campaign was strikingly different either from what had gone before (in terms of health education) or from the underlying hopes its staff held for it. In speeches to community groups, Vaughan was careful to stress the non-assimilative nature of the interventions promoted (but not supplied) by the campaign. In a briefing paper for an interview on an Asian community television programme, for example, the Minister's text read, 'Won't this interfere with traditional Asian ways of living? No. It would be quite wrong to advise people to eat foods they would not normally want to eat. And no-one is suggesting that people should dress in a way they regard as immodest in order to obtain sunlight'.[66] Even so, as the campaign wound down in 1982, some researchers and many DHSS staff still commonly assumed a link between westernisation and health, distinctiveness and disease, after the campaign.[67] The idea of assimilation as the long-term solution to 'Asian rickets' persisted even in the face of contrasting evidence – elicited by the campaign and the new lines of communication it established – from the communities themselves. Members of Britain's diverse Asian ethnic communities had presented the Department with a different picture of the causes of 'Asian rickets', one which familiar environmental understandings of rickets as a disease of poverty and poor housing could encompass more easily than explanations tailored around assumptions of exotic difference. These, in turn, illustrated the ways in which such assumptions, held on a societal level, might themselves be pathogenic.[68]

Yet perhaps the most striking evidence of the limits of both the campaign and the new 'public health' it epitomised comes from an off-the-cuff comment in the *BMJ*'s glowing report: 'Most people are probably unaware that it has been taking place because it has been aimed at the immigrant population, particularly the Asians, among whom rickets is prevalent'. 'Stop Rickets' was not then a 'public health campaign' in the sense established by the mass campaigns against TB or nutritional deficiency during and immediately after the Second World War. It was, in fact, invisible to 'most people'. The *BMJ* continued:

> Of all the immigrant communities, the Asians tend to be the most isolated, partly because they cling to their traditional social habits... Rickets... to most people is a thing of the past... The trouble is that a diet suitable to the Indian subcontinent simply does not contain the vitamins and oils needed in Britain's climate, and the campaign helps people understand what changes they must make.[69]

In contrast to the Ministry's earlier responses to TB among Indian, Pakistani, East African and Bangladeshi migrants, Britain's now-established Asian-origin populations had become visible to the central health authorities as a legitimate focus for targeted health campaigns and educational, acculturating interventions. However, they remained distinct from 'the public' and were still easily figured as an 'immigrant' group. In the end, the onus remained on these communities to change, in fundamental ways, in order to earn both health and directly responsive health care under grey British skies.

Conclusion

In post-war Britain, health – and perhaps population health, even more than the health of individuals – became both a marker of national modernity and an indication of social inclusion. The extent to which some migrant groups were included within the broad remit of 'public

health' routine therefore indicates the degree to which their contributions (principally as a flexible low-paid workforce driving Britain's
economic reconstruction) were visible and valued and the extent to
which they were regarded as (potentially) 'hygienic citizens'.[70] Notably,
in relation to TB, the Irish fall into this category, as do the less pub-
licly visible but far more thoroughly monitored European Voluntary
Workers. Contrastingly, the fact that those migrant groups most clearly
and consistently racialised were almost totally excluded from national
visions of 'the public health' reflects the extent to which their contributions were devalued or obscured by the physical and cultural
distinctiveness which was simultaneously seen as a barrier to their
eventual assimilation into the British body politic. Their health seemingly mattered less to Britain's recovery and – controversy and moral
panic notwithstanding – their enclave-bound infections presented only
minor, local threats to the general public.

However, the exclusion of South Asian (and to a slightly lesser degree,
West Indian) migrants from professional and political understandings
of 'the public health' also reflects their very patchy distribution across
Britain, their initially low numbers, and considerable uncertainty and
anxiety within the Ministry of Health and its successors about how to
approach these culturally distinctive populations. All too often rooted
in untested assumptions and stereotypes, I would argue that these fears
were nonetheless genuinely felt. Indeed, specific central guidance was
eventually required to clarify the legality even of collecting health data
according to ethnicity or national origin. Only in the 1980s, with the
demographic shift from 'immigrants' to British-born 'minority ethnic
groups' and as the Race Relations law finally gained regulatory teeth
did racialised ethnic populations gradually come within the remit of
public health. By then, as the 'Stop Rickets' campaign illustrates and as
John Welshman outlines elsewhere in this volume, many of the principles and assumptions which had underpinned earlier models of public
health had themselves begun to change.

Notes

1. See Anne Hardy, 'Reframing Disease: Changing Perceptions of Tuberculosis
   in England and Wales, 1938–70', *Historical Research*, 76 (2003), 535–56.
2. See Rima Apple, 'Vitamins Win the War: Nutrition, Commerce and Patriotism in the United States during the Second World War', in David Smith and
   Jim Phillips (eds), *Food, Science Policy, and Regulation in the Twentieth Century: International and Comparative Perspectives* (London: Routledge, 2000),
   135–49; David F. Smith, 'Nutrition Science and the Two World Wars', in

David F. Smith (ed.), *Nutrition in Britain: Science, Scientists and Politics in the Twentieth Century* (London: Routledge, 1997), 142–66.

3. MMR was a pre-war innovation that by the 1950s allowed mass chest screening through the production of small-scale photographs of x-ray fluoroscope screens operated either from static units or mobile radiography vans.

4. Hardy, 'Reframing Disease'; Linda Bryder, *Below the Magic Mountain: A Social History of Tuberculosis in Twentieth-Century Britain* (Oxford: Oxford University Press, 1988), 227; John Welshman, *Municipal Medicine: Public Health in Twentieth Century Britain* (Oxford: Peter Lang, 2000), 149–57.

5. The National Archive, Kew (TNA) MH55/2275 Daniel Thomson to Michael Reed, 29 October 1955.

6. 'Glasgow's M.M.R. Campaign', *Lancet*, 272 (8 November 1958), 1004. See also Ian Levitt, 'TB, Glasgow and the Mass Radiography Campaign in the Nineteen Fifties: A Democratic Health Service in Action', Scottish Health History: International Contexts, Contemporary Perspectives (20 June 2003) www.gla.ac.uk/media/media_148456_en.pdf [accessed 01/06/2011].

7. J.E. Geddes, 'Tuberculosis To-day', *Public Health*, 74 (December 1959), 96–111, 109.

8. Burrell Collection Photo Library Ref. 930.99.142 available online at www.theglasgowstory.com/image.php?inum=TGSE00889 [accessed 01/06/2012].

9. 'Glasgow's M.M.R. Campaign'.

10. James Cuthbert, 'Bronchogenic Carcinoma: A Mass Radiography Group Compared with a Practitioners' Group', *British Journal of Diseases of the Chest*, 53 (July 1959), 217–25.

11. H.E. Seiler, A.G. Welstead and J. Williamson, 'Report on Edinburgh X-ray Campaign, 1958: Introduction and Community Arrangements Technical and Administrative Arrangements Results of Campaign', *Tubercle*, 39 (December 1958), 339–59.

12. 'Glasgow's M.M.R. Campaign'.

13. TNA MH55/2276 Clipping, 'The Disease People Thought Was Beaten', *News Chronicle*, 5 August 1960.

14. Geddes, 'Tuberculosis To-day', 110.

15. 'Liverpool Mass-Radiography Campaign', *Lancet*, 275 (28 May 1960), 1180.

16. Seiler et al., 'Edinburgh, 1958', 359.

17. See Cox, Marland and York's chapter in this volume for attitudes towards Irish migrants in nineteenth-century England.

18. On Irish 'invisibility', see Liam Greenslade, Moss Madden and Maggie Pearson, 'From Visible to Invisible: The "Problem" of the Health of Irish People in Britain', in Lara Marks and Michael Worboys (eds), *Migrants, Minorities and Health: Historical and Contemporary Studies* (London and New York: Routledge, 1997), 147–78 and Anne Mac Lellan's chapter in this volume. The overall Irish rate of TB infection was 230 per 100,000 in 1952 but there were pronounced differences between rates of infection among 'tuberculinised' urban and immunologically naïve rural populations.

19. TNA MH55/2275, Annotation, Thomson to Prideaux, 28 October 1955.

20. David A. Coleman, 'U.K. Statistics on Immigration: Development and Limitations', *International Migration Review*, 21 (Winter 1987), 1138–69, 1143.

21. Oscar Gish, 'Color and Skill: British Immigration, 1955–1968', *International Migration Review*, 3 (Autumn 1968), 19–37.

22. 'Irish Immigrants (Tuberculosis)', *Hansard* House of Commons Debate (HC Deb.), 544 cols. 806–8 (25 July 1955).

23. Barnett Stross, 'Immigrants (Tuberculosis)', *Hansard* HC Deb., 544 cols. 1,308–18 (27 July 1955), on col. 1,308.

24. See John Welshman, 'Importation, Deprivation, and Susceptibility: Tuberculosis Narratives in Postwar Britain', in Flurin Condrau and Michael Worboys (eds), *Tuberculosis Then and Now: Perspectives on the History of an Infectious Disease* (Montreal: McGill-Queen's University Press, 2010), 123–47; John Welshman, 'Compulsion, Localism, and Pragmatism: The Micro-Politics of Tuberculosis Screening in the United Kingdom, 1950–1965', *Social History of Medicine*, 16 (2006), 295–312; John Welshman and Alison Bashford, 'Tuberculosis, Migration, and Medical Examination: Lessons from History', *Journal of Epidemiology and Community Health*, 60 (2006), 282–4; John Welshman, 'Tuberculosis, "Race", and Migration, 1950–70', *Medical Historian: Bulletin of Liverpool Medical History Society*, 15 (2003–04), 36–53; John Welshman, 'Tuberculosis and Ethnicity in England and Wales, 1950–70', *Sociology of Health and Illness*, 26 (2000), 858–82; Ian Convery, John Welshman and Alison Bashford, 'Where is the Border? Screening for Tuberculosis in the United Kingdom and Australia, 1950–2000', in Alison Bashford (ed.), *Medicine at the Border: Disease, Globalization and Security, 1850 to the Present* (Houndmills: Palgrave Macmillan, 2006), 97–115.

25. See, for example, TNA MH58/670 Letter, John Fishwick (Town Clerk, Borough of Lambeth) to Ministry of Health, 28 February 1955.

26. Patricia Hornsby-Smith, 'Immigrants (Tuberculosis)', *Hansard* HC Deb., 544 cols. 1,308–18 (27 July 1955), on col. 1,314.

27. Ibid.

28. See Wendy Webster, *Imagining Home: Gender, 'Race' and National Identity, 1945–64* (London: University College London Press, 1998), 91–127.

29. Hornsby-Smith, 'Immigrants (Tuberculosis)', col. 1,318.

30. TNA MH148/30, 'Minister's Meeting with British Medical Association', 3 November 1964.

31. See Welshman, 'Tuberculosis Narratives', 128 and TNA MH58/60, TNA MH148/30.

32. TNA MH55/2275, Memo, E.T. Prideaux to M. Reed, 1 February 1955; TNA MH55/2275, E.T. Prideaux, 'Social and Economic Problems Arising from the Growing Influx into the United Kingdom of Coloured Workers from other Commonwealth Countries', 11 July 1955.

33. Hornsby-Smith, 'Immigrants (Tuberculosis)', col. 1,318.

34. Sheila Tobias, 'The Ostrich Game', *The Listener*, 15 April 1965. n.p.

35. See Anne Murcott, 'Food and Nutrition in Post-War Britain', in James Obelkevich and Peter Catterall (eds), *Understanding Post-War British Society* (London: Routledge, 1994), 155–64.

36. See Charles Webster, 'Government Policy on School Meals and Welfare Foods, 1939–1970', in Smith (ed.), *Nutrition in Britain*, 190–213; Ministry of Health and Department of Health for Scotland, 'Report of the Joint Sub-Committee on Welfare Foods' (London: HMSO, 1957) and W.T.C. Berry, 'Nutritional Aspects of Food Policy', *Proceedings of the Nutrition Society*, 27 (1968), 1–8.

37. On the US example, see Apple, 'Vitamins Win the War'; Rima Apple, '"If I don't get my vitamins...anything can happen": Nutrition in Twentieth Century Popular Culture', *Pharmacy in History*, 40 (1998), 123–31, and for milk in particular, Rima Apple, *Vitamania: Vitamins in American Culture* (Rutgers, NJ: Rutgers University Press, 1996), 33–53. For Britain, see Smith, *Nutrition in Britain*; Harmke Kamminga and Andrew Cunningham (eds), *The Science and Culture of Nutrition, 1840–1940* (Amsterdam: Rodopi, 1995).

38. See A. Neuberger, 'Charles Enrique Dent', *Biographical Memoirs of Fellows of the Royal Society*, 24 (1978), 14–31 for a sense of the trans-Atlantic excitement about biochemistry as it played out in an individual career path. For a historical overview, see Robert E. Kohler, *From Medical Chemistry to Biochemistry: The Making of a Biomedical Discipline* (Cambridge: Cambridge University Press, 1982).

39. Wellcome Trust Contemporary Medical Archives (CMAC), PP/CED/C.3/1/3, C.E. Dent to J. Kyle Smith, 10 September 1960.

40. See Roberta Bivins, '"The English Disease" or "Asian Rickets"? Medical Responses to Postcolonial Immigration', *Bulletin of the History of Medicine*, 81 (2007), 533–68, for a detailed analysis of this aspect. Compare too with the ways in which the Navajo at Many Farms in the USA served US antibiotics researchers as a 'Third World country in the USA': David S. Jones, 'The Health Care Experiments at Many Farms: The Navajo, Tuberculosis, and the Limits of Modern Medicine, 1952–1962', *Bulletin of the History of Medicine*, 76 (2002), 749–90.

41. TNA MH148/623, S.W. Stanbury to J.M.L. Stephen, 6 March 1973.

42. Berry, 'Nutritional Aspects', 3.

43. Ibid., 7.

44. See, for example, Lord Aberdare, 'Parliamentary Question (PQ) 98/72/73', *Hansard* House of Lords (HL) 340, cols. 300–314 (14 March 1973).

45. Ernest Fernyhough, 'Existing Development Areas – England and Wales', *Hansard* HC Deb., 821, cols. 519–77 (14 July 197).

46. Jimmy Hamilton, 'Existing Development Areas – England and Wales', *Hansard* HC Deb., 821, cols. 519–77 (14 July 1971), on 563–4. Hamilton was MP for Bothwell, then a relatively affluent suburb of Glasgow.

47. 'Rickets', *Hansard* HC Deb., 850, cols. 1,134–5 (13 February 1973).

48. Sylvia J. Darke, 'Monitoring the Nutritional Status of the UK Population', *Proceedings of the Nutrition Society*, 36 (1977), 235–40, on 240 (emphasis added).

49. Darke, 'Monitoring', 236.

50. See Bivins, '"English Disease" or "Asian Rickets"', 559; TNA MH148/624 Memo, Darke to Dr A. Yarrow, 13 December 1978.

51. Darke, 'Monitoring', 239; TNA MH148/624 Memo, Yarrow to Darke, 21 December 1978.

52. Amanda Engineer, 'The Society of Medical Officers of Health: Its History and its Archive', *Medical History*, 45 (2001), 97–114; Ann Oakley, 'Making Medicine Social: The Case of the Two Dogs with Bent Legs', in Dorothy Porter (ed.), *Social Medicine and Medical Sociology in the Twentieth Century* (Amsterdam: Rodopi, 1997), 81–96.

53. TNA MH148/623, Draft Pamphlet, 'Vitamin D Deficiency – Rickets and Osteomalacia', COMA (Nutrition) Panel on Child Nutrition, January 1976.
54. TNA MH148/623, 'Vitamin D Deficiency – Rickets and Osteomalacia'.
55. It is, however, worth noting that indigenous sufferers and the 'more assimilated' West Indians were elsewhere still blamed for their children's rickets on the grounds of irresponsible mothering and 'food faddism'.
56. TNA MH148/623, Minutes, COMA (Nutrition) Panel on Child Nutrition Meeting, 16 July 1975.
57. See, for example, TNA MH148/624, W.T. Cooke to S.J. Darke, 5 May 1977; TNA MH148/624, J.L.H. O'Riordan to E.M. Widdowson (Chairman of COMA, Dunn Nutritional Laboratory), 31 July 1978; Derek Humphrey, 'MP Seeks to Curb Rickets', *Sunday Times*, 20 November 1977.
58. 'Aid Urged for Asians affected by Rickets', *Guardian*, 5 July 1979.
59. See Keith Wailoo, *Dying in the City of the Blues: Sickle Cell Anaemia and the Politics of Race and Health* (Chapel Hill, NC: University of North Carolina Press, 2001), especially Chapter 6.
60. TNA MH160/1449, Minute, T.M. Gang to Wormald, 9 April 1980.
61. TNA MH160/1447, 'Asians: Rickets – Meeting with Dr Vaughan, Minister of State (Health) 31 July 1979'.
62. TNA MH160/1449, Memo, T.M. Gang, 'Campaign on Rickets – Appointment of Director', 18 April 1980. For her perspective on the campaign, see Veena Bahl, 'Stop Rickets Campaign', *Nutrition & Food Science*, 81 (1993), 2–5.
63. William Russell, 'Letter From Westminster: Smiles About Rickets, Gloom Over Whooping Cough', *British Medical Journal*, 284 (30 January 1982), 358.
64. Russell, 'Smiles About Rickets', 358.
65. Hedley Peach, 'A Review of Aetiological and Intervention Studies on Rickets and Osteomalacia in the United Kingdom', *Community Medicine*, 6 (1984), 119–26.
66. TNA MH160/1451, Briefing Notes, MS(H) 'Nai Zindagi Naya Jeevan' TV interview 12 February 1981.
67. W.P. Stephens, P.S. Klimiuk, S. Warrington, J.L. Taylor, J.L. Berry and E.B. Mawer, 'Observations on the Natural History of Vitamin D Deficiency Amongst Asian Immigrants', *Quarterly Journal of Medicine*, 51 (1982), 171–88.
68. See, for example, TNA MH160/1451, Letter, Dr Qudratul Islam to Dr Gerald Vaughan, 10 March 1981.
69. Russell, 'Smiles About Rickets', 358.
70. For more on 'hygienic citizenship', see Warwick Anderson, *Colonial Pathologies: American Tropical Medicine, Race, and Hygiene in the Philippines* (Durham, NC: Duke University Press, 2006); Alison Bashford, *Imperial Hygiene: A Critical History of Colonialism, Nationalism and Public Health* (Houndmills: Palgrave Macmillan, 2004).

7
Immigration and Body Politic: Vaccination Policy and Practices during Mass Immigration to Israel (1948–1956)

Nadav Davidovitch

Though vaccinations are usually considered a paradigm of bio-medical success, their use has frequently provoked fierce criticism and unparalleled opposition.[1] This article focuses on the social history of vaccination policy and practices during the period of mass immigration to Israel. Between 1948 and 1956, the newly established country, with a population of only about 700,000, faced the formidable task of absorbing over one million new immigrants. Following a short overview of health and immigration policies during the first years of the newly established Israeli State, this chapter will focus on vaccination as a case study to demonstrate the reciprocal relationships between the health system, various health agents and the immigrants – particularly the immigrant's body as an entity that the state seeks to supervise and define. Although the Israeli vaccination programme for immigrants was generally described by its designers as an unproblematic and necessary step in transforming the immigrants into members of 'modern civilisation', deeper research reveals that on many occasions vaccination policy did indeed encounter difficulties. As historian David Arnold has claimed, states supervise and control the body politic by disciplining individuals' bodies.[2] Vaccinations, as part of a broader system of regulations that govern the care of infants, hygiene and health, constitute one example of the ways countries supervise the bodies of their citizens. Historians have recognised that opposition to vaccination can help us understand the politics of the body and its relation to the modern state.[3] The use of state authority and power in implementing public health measures is all the more amplified when it is applied to marginalised populations, often consisting of ethnic minorities and migrants.

Immigrants who underwent vaccination and other public health interventions were not 'passive raw material' to be moulded by various authorities. They possessed a world outlook on health and illness that was not always in keeping with the plans of the absorbing society. This article explores the complex interactions between public health authorities, operating in the name of the state, health care workers who implemented various public health campaigns, and various groups of immigrants entering Israel between 1948 and 1956. These interactions between public health personnel and immigrants are not peculiar to the Israeli case. However, in recent years it has become evident from a large number of studies that local context plays a major role. Thus, one cannot simply learn from the experiences of other case studies.[4] Vaccination policies and practices in the *Ma'abarot* (transit camps),[5] generally not interrogated in the various historical accounts of mass immigration to Israel, constitute a thick case study for understanding the interaction between absorbing systems and immigrants in terms of conflicting perceptions of illness and health.

Zionism, health and immigration

Zionist ideology championed an 'Ingathering of the Exiles' as one of the most important objectives of the Zionist movement, and later of the State of Israel. The use of the value-loaded term *aliyah* – meaning 'to climb' – rather than the neutral term in Hebrew applying to immigration anywhere, *hagira,* is indicative of the different light in which Zionism views immigration to Israel. Yet, over the years, tensions have arisen at times between the outlook that perceives immigrants or *olim* as the most important asset of the nation and the depiction of newcomers as *chomer enoshi* (human material), a term that reflects both fears and suspicion of changes that immigrants were liable to impose on veteran Jewish society.

Immediately after its founding, the State of Israel faced the tremendous challenge of mass immigration. Within a few years the Jewish population of the state had to absorb hundreds of thousands of immigrants, many of them ill. Common ailments that required immediate attention included tuberculosis (TB), trachoma and ringworm.[6] Despite the fact that Zionist ideology viewed the new state as the homeland of the Jewish People and championed the 'Ingathering of the Exiles', in practice many apprehensions were voiced about mass immigration leading to loss of control over the character of the Zionist endeavour. It was often said that the 'human material', including Holocaust survivors and immigrants from Asiatic and North African countries, was problematic: could immigrants withstand the burden of building a new nation without turning into a burden themselves? The concept of the 'melting pot' helped to overcome these problems; a new Israeli national identity was forged out of the perceived blend of new immigrants.[7] Immigrants, particularly those from Arab countries, were depicted as possessing 'a primitive soul'. This characterisation was endemic and can be found in newspaper reportage, in various debates in the Knesset and in scientific writings (sociological, psychological and medical) of the period.[8]

The health system constituted a key component for the absorption of immigrants and for melting pot policy. Various health officials called for a preventive medicine scheme and a health education network that would go beyond the immediate treatment of various diseases. A complex system of welfare workers, nurses and doctors worked together, striving not only to heal immigrants but also to educate them in a host of areas, from infant care to matters of personal hygiene. These systems did not arise in a vacuum; in the 1920s and 1930s, the *Yishuv* had developed wide-ranging 'recommended' hygienic practices as an inherent part of Zionist ideology. This activity, developed by health institutions such as Hadassah and the Labor Federation's General Sick Fund, worked in close cooperation with various organisations in the realm of education and immigrant absorption. These programmes drew on wellsprings of American and European influences, but also arose, at least in part, from the Zionist establishment's conception of creating a 'new man'.[9] This project was similar in many ways to projects carried out in Europe and the United States at the time and eugenic influences are apparent.[10] The eugenic outlook intertwined with colonialist practices and presented the white/European body as the 'right' model.[11] The Zionist movement with its European foundations, and aspirations to forge a 'New Jew', fitted in well with this approach. Public health policy towards immigrants in Israel was founded on a similar belief in public health practices as a vehicle for moulding a person who was healthy in body and soul. This was deemed to be an important goal not only at the level of the health of the individual, but also at the level of the nation as a whole. While these public health campaigns brought about a significant improvement in health indices, such as rates of infant and maternal mortality and the incidences of and mortality from infectious diseases, they also carried with them a complex social cost.

While the attempts to liberate the new society from the 'chains of the diaspora Jew' were fundamental to Zionist thought from its

Table 7.1   Immigrants to Israel according to country of origin, 1948–1954

Year/Origin	Europe-America	Asia	Africa	Unknown	Total
1948	75.5%	4.5%	8.0%	12.0%	100% 101,819
1949	51.5%	30.0%	16.5%	2.0%	100% 239,076
1950	50.5%	34.0%	15.0%	1.0%	100% 169,405
1951	29.0%	59.0%	11.5%	0.5%	100% 173,901
1952	30.0%	28.5%	41.5%	–	100% 23,357
1953	28.0%	27.5%	44.0%	0.5%	100% 10,347
1954	14.0%	18.5%	67.0%	0.5%	100% 17,471
Total Number of People	343,949	251,279	120,752	19,396	735,376

Source: Israeli Central Bureau of Statistics, Statistical Abstract of Israel (2011), Table 4.2, 230.

inception, new dimensions were added to this goal, with the change in the demography of immigration to Israel in the early 1950s as migration of Jews from Arab countries came to outweigh migration from Europe (mainly Holocaust survivors) (see Table 7.1).

This demographic change merits special attention in understanding later frictions between newcomers and the host society. Before the Holocaust, the main divide within European Jewish society was between the Western European Jews and Eastern European Jews, the *Ostjuden*. The *Ostjuden*, who lived in great poverty and suffered from malnutrition and a variety of epidemic diseases, were the target of public health campaigns both in Europe and in other countries to which they migrated, such as the United States and Palestine during the British Mandate. The Zionist movement also perceived these Jews as the main human reservoir that should be approached when potential immigrants to Palestine were sought. All this changed after the Holocaust and the establishment of Israel.

While during the British Mandate the main 'high risk' target of intervention was the 'dirty' *Ostjuden*, after the establishment of Israel the public health effort focused on the 'education' of the increasing numbers of Jews arriving from Arab countries. In addition, the need to move away from the 'Oriental' or 'Levantine' context of Israel as a country situated in the Middle East, among hostile Arab countries, played an important role in the 'need' to reform – partly by means of public health practices – the 'Arab Jews' or '*Mizrahim*'.[12]

Another indication of the centrality assigned to health matters in the absorption of migrants can be found in the wording of the Law of Return of 1950, which enjoyed the support of all of the Israeli political parties and was viewed by many as setting forth the 'character and special mission of the State of Israel as a state that carries the vision of the Redemption of Israel'.[13] It addressed what was perceived as an urgent public health issue: the medical selection of immigrants. In the original law, one of the limitations on unfettered immigration of Jews to Israel related to persons who 'could endanger public health'.[14] This category appears, at least in terms of positioning and rhetoric, adjacent to other 'dangerous categories' to be barred from immigration: a person who acts against the Jewish People who is liable to endanger the security of the state or a person with a criminal record who is liable to endanger public order. Indeed, in the first years after statehood, the issue of 'medical selection' was controversial and reflected the tensions between the young country's need for immigrants and the fear of the economic and social consequences of migration.[15]

The medical absorption system

In 1943, Dr Yosef Meir, then head of the General Sick Fund, formulated the medical programme for the absorption of immigrants together with various experts who were invited by David Ben-Gurion to design the 'One Million Plan' to absorb a million Jewish immigrants.[16] Soon after, in 1944, the Jewish Agency established the Immigrant Medical Service, which was managed by Hadassah, the volunteer Women's Zionist Organisation of America, until the establishment of the State in 1948. After the State was established, responsibility for running the Service became subject to fierce controversy, particularly in relation to its financing; in the end, the Ministry of Health took over its management.[17]

Many agencies were involved in the medical treatment of immigrants in the first years of the state, creating overlapping roles and disputes over authority. The list of bodies dealing with health included the Ministry of Health, the Jewish Agency, the American Joint Distribution Committee (AJDC), the World Zionist Organization (WZO), the General Sick Fund and a host of local organisations. Provision of medical services became a political issue as well, for membership of the General Sick Fund was closely tied to political party membership. New immigrants automatically received three months coverage exclusively through the Fund. Afterwards, they were free to choose which of the various sick funds they wished to join, but in practice many immigrants retained membership of the General Sick Fund.[18]

In fact, immigrants were already concentrated in camps in their places of origin before leaving for Israel. The objectives of these temporary processing camps (*machanot maavar*) included medical classification. The medical and epidemiological information gathered in camps was supposed to enable the medical establishment in Israel to anticipate medical problems. However, the opportunity for medical checkups or treatment was often not realised. The immigrants themselves did not view medical checkups in a positive light. For some Holocaust survivors, medical examinations triggered memories of medical 'selections' in concentration camps. Some immigrants tried to avoid medical examinations, sending others to be examined in their stead out of fear that their immigration would be delayed should a medical problem be discovered.[19] These gaps in medical screening and categorisation resulted in the adoption of a policy by which additional medical inspection and classification were carried out in the *Shaar Ha'aliyah* (Gate of immigration) processing camp upon arrival in Israel.[20] *Shaar Ha'aliyah* was established in 1949 on the foundations of an old British army camp, south of Haifa. Between 1949 and 1952, 700,000 immigrants passed through the facility and 400,000 underwent medical examinations.[21] The Ministry of Health underwrote the cost of the checkups and the medical care of immigrants within *Shaar Ha'aliyah*. Within days of arrival, every immigrant underwent a physical checkup, including examination by a dermatologist (with emphasis on diagnosis of venereal diseases, leprosy and ringworm) and an ophthalmologist (with emphasis on diagnosis of trachoma and other contagious eye ailments).[22] The General Sick Fund was made responsible for some of the immigrants' medical tests upon their arrival in the country, including screening for syphilis, as part of the process of registering for the sick fund.[23] In 1952, a centre for the treatment of children with ringworm was opened in the camp as part of a national campaign for mass ringworm irradiation.[24] One can only imagine the sense of anxiety, the overcrowding and arduous conditions in the camp. Occasionally the tension and the anger spilled over into violent outbursts towards staff, forcing physicians and medical staff to work under police protection. This was the climate in which mass vaccination against various scourges was carried out.[25]

Vaccination policy

The immigrant camps established abroad offered the first opportunity to engage in public health work, including vaccinations. According to letters and reports sent by camp physicians, there was a shortage of vaccines. There were various logistical problems resulting from the distance of the camps from supplies, as well as the difficulties at times in transporting vaccines under suitable conditions. These issues delayed full implementation of the planned vaccination programme. In addition staff had to deal with cases of side effects from vaccinations, particularly local skin infections.[26]

The legal status of the vaccination was also problematic. There was no legal basis for forcing immigrants to agree to be vaccinated when the country of origin did not require a specific vaccination.[27] In Israel smallpox vaccinations were required by law, a practice going back to the British Mandate.[28] Other vaccinations, for instance against typhoid and TB, were not obligatory, yet one may assume that all vaccinations were presented to the immigrants as important.[29] The vaccinations were administered free during the immigrants' first year in the country.[30]

Although smallpox vaccinations were given to all inhabitants of Israel – newcomers and veterans – scrutiny of contemporary records indicate that most concern focused on the administration of vaccinations and achieving a high level of coverage among immigrants. Ideally, the immigrants were to be vaccinated during their sojourn in 'transit camps' prior to embarkation for Israel, but in some cases vaccinations were carried out on the ships bound for Israel. In many cases, due to loss of records or shortage of vaccine, vaccination took place only on arrival in Israel.[31]

According to Ministry of Health policy during these years, infants were vaccinated against smallpox when they were three months old and again prior to entering the school system.[32] The technique for vaccination against smallpox at this time was not unified; various means of introducing the vaccine were used, ranging from a single prick to a series of pricks and peeling off a small piece of skin to introduce the vaccine. These methods sometimes caused infection or produced large, ugly scars.[33] At times recipients accidently transferred vaccine material to another part of the body, causing scarring, and in rare cases blindness, when children inadvertently rubbed the material in their eyes. In all cases, the sign that the vaccine had 'taken' was the appearance of a scab on the site of the vaccination a week after administration.

Consequently, children who had received vaccines had to be brought by their mothers to have the scar examined. Testimonies indicate that the scarring process was considered ugly, particularly for girls.[34] It was forbidden to bandage the vaccination area and mothers were instructed not to wash their children for three days after vaccination. The vaccination area had to be kept dry for a week. At the end of the seven-day period the child was examined to see whether the vaccine had taken. If not, the child had to be re-vaccinated.[35] In the years 1949, 1950 and 1953, following word of a smallpox outbreak in neighbouring countries, mass vaccination campaigns were announced in the press, on the radio and street posters, urging individuals to seek vaccination.[36]

Vaccination against typhoid was already common for tourists and immigrants during the British Mandate period. Typhoid vaccination did not, however, share the same obligatory statutory status and the entire population was not targeted. Instead campaigns focused primarily on new immigrants and soldiers.[37] The utility of typhoid vaccination was subject to controversy. Some physicians argued that it could not replace good hygienic practices that were inculcated in various public health projects; they felt vaccination, in essence, might undercut such campaigns. This was not the case with smallpox, a contagious disease that was not perceived to be as closely tied to personal hygiene.

Despite the debate over its utility, vaccination of immigrants against typhoid was viewed as important and remained a component of the vaccination programme until the mid-1950s.[38] A series of three vaccinations was required to achieve full immunity and, under the conditions prevailing in Israel in the 1950s, the procedure was difficult to administer in terms of follow-up. Vaccinations could not be fully administered outside Israel and it was problematic to complete the series on time after arrival in the country. The reports of epidemiologists estimated that coverage was 55 per cent, a figure that was considered below herd immunity. This failed to provide full protection to the population against an outbreak of typhoid, a fact that made the vaccination programme even more controversial. In 1954 the Epidemiology Department of the Ministry of Health recommended that only school-age children be vaccinated, and in 1957 typhoid vaccination was also removed from school vaccination programmes.[39]

BCG vaccination against TB was also administered on a large scale. TB constituted one of the primary health problems facing the health system in the early years of statehood.[40] The numerous immigrants with TB who arrived in the country were viewed as a heavy burden on the health system. Dr Yosef Meir, Director General of the Ministry of Health in 1949, recommended that due to the severity of the problem and the shortage of hospital beds that the immigration of people suffering from TB should be delayed until individuals had been cured abroad, with AJDC and the WZO covering the cost of care.[41] The number of TB patients was extraordinary high among immigrants from Eastern Europe, particularly Holocaust survivors.

Due to the prevalence of the disease, all immigrants were supposed to undergo Mantoux tests to establish exposure to TB and the need for vaccination. Between November 1949 and November 1950, a large TB vaccination campaign was launched that included preliminary screening and vaccination of populations who had been exposed to the disease. The campaign, which was carried out with the assistance of the Danish and Swedish Red Cross and a Norwegian relief organisation, encompassed the screening of 365,298 individuals and the vaccination of 208,851 of those who had been screened.[42]

Negotiation and resistance

So far, this account has focused on the cut-and-dry 'official story' as recorded in various Ministry of Health reports.[43] Yet, a deeper examination based on additional documentation and written and oral testimonies reveals a far more complex reality. A bleak picture emerges from the reports of doctors and nurses who visited the immigrant camps. Sanitary conditions were dismal; piles of trash and open sewage were common.[44] Poor sanitation and other adverse conditions were reflected in the statistics: infant mortality spiralled, peaking in 1950 at 157 deaths per thousand live births among residents of the Ma'abarot. All these facts generated harsh criticism.[45]

Public health officials also encountered grave difficulties in implementing preventive health policies, including vaccination. Despite the fact that vaccination was supposed to be documented in the immigrants' registration cards and in clinic logs, in practice records were not always accurate and sometimes documents were lost, leading to needless re-vaccination. Dr Chaim Sheba, Deputy General of the Ministry of Health in 1951, underscored the severity of the situation; in a meeting with the Coordination Institute on 11 July 1951, he suggested that receipt of food ration stamps be made conditional on immigrants proving that they had been vaccinated as required:

> In the campaign to bring in Iraqi Jews, certain concessions had to be made on the tight ring of monitoring...The immigrants have been incited and don't allow Ministry of Health personnel to carry out even one test or inject one vaccination, and the suggestion on our

part to ensure at least injection of vaccination by restrictions on food cards was not accepted by the Jewish Agency people, out of fear of the wrath of the immigrants.

The result is that there is no vaccination against typhoid. What are the ramifications? That we have an incidence rate 50 times higher of this malady than what has been achieved in countries with orderly hygiene and we could have come close to this number through vaccination.[46]

The subtext of Sheba's complaints was the change in the demography of the immigrants, discussed above. The immigrants referred to in the text are Iraqi Jews who according to Sheba were resisting typhoid vaccination. It is interesting to closely read Sheba's argument about the need to compare high typhus rates (caused by the arrival of the immigrants) to those in 'countries with orderly hygiene' and the medicalised solution of vaccination, including the suggestion to restrict access to food, while neglecting the harsh social and hygienic conditions in the transit camps.

Sheba was not alone in suggesting measures to increase vaccination rates. In the first conference of paediatricians held in Natanya, a resolution called upon the Ministry of Education to demand that every child entering kindergarten present a document certifying that they had received a diphtheria shot.[47] Solutions such as this attempted to tie preventive medicine – whether through vaccination or medication – to receipt of another 'service' and were apparently considered legitimate in the eyes of many doctors. According to Dr Abraham Sternberg, head of the Immigrant Medical Service, a similar strategy of making receipt of food stamps conditional upon acceptance of preventive medical measures was instituted in the transit camp in Aden, an out-of-sight location that was liable to generate less public debate:

In the camp compound we had no other alternative but to tread a cruel path... There was no point in trying to explain the danger of the malaria from which most of the immigrants suffered. We would not have succeeded in convincing them to come willingly of their own volition to accept treatment... It's questionable whether they had ever seen medicine in pill form before... We decided therefore to tie receipt of food to each family on acceptance of medication against malaria... we felt that in this manner of carrying out [our mission], we were saving souls. But the immigrants didn't know, of course, and

couldn't understand what the meaning was of this terrible spectacle to which they were innocently subjected, crowded together in
masses... swallowing pills under the gaze of strict wide-eyed supervi-
sors... Day-after-day 5,000 souls went through the big tent. We began
at six in the morning and at three in the afternoon the last families
received the stamp on the food card and only then could they receive
and prepare their daily food.[48]

The ambivalence of the author, who clearly sensed the tension between
health personnel and immigrants, was justified by Dr Sternberg's asser-
tion that he and his colleagues 'were saving souls'. But he remained
convinced that 'the immigrants, don't know, of course, and can't
understand what's the meaning of this terrible spectacle'.

The same attitude was reflected in the impressions of a nurse who
worked in the *Ma'abarot* in the 1950s and who spoke of the 'inability'
to explain to the immigrants what was going on and why:

The vaccinations were not known to the majority of immigrants and
their objection to vaccination was strident at times and a source of
great frustration to the nurse. One could compare the attempt to
explain and convince the immigrants of the need for vaccination to
an attempt to explain to an infant the meaning of treatment with
DDT or vaccination against smallpox.[49]

This motif, that there was no sense in explaining to newcomers the
meaning of various medical procedures, due to their mental and spir-
itual 'limitations', is repeated time and again by medical staff of the
period. The logic behind this perception was that the medical personnel, whether doctors or nurses or any other emissaries of the medical
establishment, could make decisions on health issues in light of the
preferential knowledge they held. Of course this knowledge was strongly
anchored to a hegemonic perception about the ways in which the immigrant's body should be made healthy and integrated into the body
politic of the new state. The 'wardenship' over the bodies of inhabitants of Israel in the 1950s, one of the peak points of nation-building,
fits into what anthropologist Meira Weiss labels 'the culture of the Cho-
sen Body', the very culture that sought to create an Israeli collective
identity.[50] But these perceptions did not always sit well with the perception of the body and the health and illness of those being absorbed, and
resulted in tensions and clashes.

In response to the official public health policy expressed by Sheba, Sternberg and public health nurses, immigrants developed their own public health perceptions. It did not necessarily mean automatic rejection or resistance. Most of the time immigrants complied with and accepted public health measures in the belief that these practices would improve their health or would lead to a quicker assimilation into Israeli society. Yet, from time to time, public health practices provoked strong resistance. Mass vaccination, treatment for malaria or ringworm and other preventive medicine measures could trigger opposition and contempt. Usually these events were unremarkable. They took place in a grey and harsh daily regime or routine that is difficult to reconstruct. This 'micro-resistance' that had surfaced mostly on a local level pointed to the interaction and conflicts between public health personnel and immigrants. The newcomers had space – small and problematic sometimes – to manoeuvre and negotiate. Of course not all immigrants reacted in the same manner and they cannot be regarded as a homogeneous entity. Similarly, the agents operating in the public health realm emanated from various professions and traditions – physicians, nurses, social workers and so on, and many were immigrants themselves. Yet, a common feature of the negotiation process between immigrants and health workers was that it involved practices inflicted on the body.

Vaccination and the body

Dr Sheba's recommendations and those of the paediatricians were not adopted or put into practice in Israel, but they testify to the paternalistic-collectivist thinking that was common at the time. As mentioned above, this way of thinking was not unique to the Israeli health system. Yet, it is important not to lose sight of the local context in which most of the vaccinations were administered at the time and in fact are administered in Israel to this day.

Vaccination of children in Israel, which comprise the overwhelming majority of vaccinations, are administered in Family Health Stations[51] or – as they are still called by most Israelis – *Tipat Chalav* ('A drop of milk') clinics.[52] The 'mother and child clinics' were initially brought to Palestine in 1913 at the urging and with financial support from the Jewish philanthropist Natan Strauss. They were expanded by Hadassah under the leadership of Henrietta Szold, and further developed by Hadassah-WIZO (a Jewish American volunteer organisation) and the General Sick Fund during the period of the British Mandate, ultimately evolving into a unique Israeli institution which monitored mothers and children and constituted a core facet of hygienic work within the Zionist *Yishuv*. This institution symbolised the special place assigned to children and new mothers raising children as a cornerstone in revitalising and building the nation. Safeguarding the young child's well-being was considered a central issue that demanded investment and forethought, taking strides to ensure adherence to the codes set down by public health personnel. The figure who carried out this policy was the *Tipat Chalav* nurse – a trained public health nurse – who, in essence, ran the station. *Tipat Chalav* stations also vaccinated children, an activity that was viewed of utmost importance in safeguarding the health of the child.

The centrality of children and the correct education of mothers were even more marked during the great immigration of the 1950s. *Tipat Chalav* clinics were established throughout the country with the mission of educating mothers and monitoring the growth of young children as the country's most important asset. The *Tipat Chalav* nurse taught young mothers how to diaper, feed and bathe their offspring. In addition, the nurses carried out home visits to new immigrant mothers teaching them the details of housekeeping. The nurses kept diaries in which they recorded their impressions of these home visits. In one typical diary the author who worked in a *Tipat Chalav* clinic in a *Ma'abara* recorded:

30.1.1951: The problem and its essence:...The mothers don't know how to behave with a sick child – to give it warm sweet liquids and now to give it pills...to care for the new mother and the child, make her bed, to wash her, to give her the necessary medical care after delivery. Comments: The woman sits on a sack...The dirt is terrible.

31.1.1951: General comments: The new mother absolutely refuses to let a doctor care for her. She won't call a physician to give birth and didn't even yell in her labor so that the army would not transfer her to a hospital.[53]

These short passages demonstrate the tremendous power invested in the hands of public health nurses. Here we find evidence of descriptions of a mother, a new immigrant who is viewed as not trained in how to 'behave' with her offspring, based on her unwillingness to give birth in a hospital and her inability to feed and care for her child. But these mothers were not 'passive'. As several studies have shown, immigrants knew how to select messages and practices from among those proffered to them by nurses, social workers and other agents in the health system, adopting those that they regarded as suitable and that reflected their worldview and perceived to be in the best interests of their families.[54] From various testimonies of new immigrants in the 1950s it becomes apparent that many viewed the nurse as a useful figure and her intrusion into the private sphere as positive, and that, indeed, these public health nurses constituted central agents for change on behalf of the state. However, mothers did not always see eye to eye with nurses and often there was friction.

It is useful to consider the practice of vaccinating children as part of a larger conflict over perceptions of health and illness, care of children and the intrusion of the state into the private sphere, where the practice of vaccination was, in the final analysis, a struggle over 'the body of the child' played out through the injection of vaccine into the bodies of the young. In this wider context it is easier to understand the testimony of Phyllis Palgi, the first anthropologist to work with the Ministry of Health in 1953, who related how in a visit she conducted with public health nurses to one of the new settlements containing immigrants from the Atlas Mountains of Morocco, the immigrants threw stones at the nurses who had come to vaccinate their children. Rumour had it that the nurses had come to put 'tainted blood' in the bodies of their children.[55]

Rumours about inoculating with contaminated blood and abuse of 'medical power' against marginalised populations was not confined to health issues in the 1950s. Various anthropological studies have demonstrated that immigrant populations in other countries such as Australia, Canada and the United States witnessed the spread of similar rumours, a phenomenon labelled 'medical gossip'.[56] According to Manderson and Allotey, medical gossip among immigrant groups serves a number of objectives. In the short term, such rumours create conflicts with the local health system. In the long run, through various cultural facilitators who come in contact with immigrant populations on the one hand and the medical system on the other, the system internalises the criticism and attempts to improve communication between staff and immigrants. Thus, such tension can ultimately lead to better understanding. However, it is important to keep in mind that the tensions reflect power relationships and attempts by the state to model the bodies of immigrants. Often, the struggle is carried out vis-à-vis control over the bodies of children.

The control of the state over the bodies of children was also manifested in a more significant and pervasive manner. At a time when immigrant encampments comprised 'tent cities', preceding the Ma'abarot of 'enhanced housing' in wooden shacks and tin huts, all infants were transferred to children's houses. This area was almost 'off limits' to parents, and the babies were left under the exclusive care of the children's house personnel. Through strict enforcement of limited visiting rights to feeding and nursing times, a state machinery holding 'mastery over the bodies of children' was created;[57] the same children symbolised very different things to their parents and the medical teams. There were cases where parents went to the children's house, only to be told that their infant had been transferred to a hospital, based on the evaluation of the medical staff, a decision taken without consulting the parents. To travel from these tent encampments or from a *Ma'abara* to a hospital was an extremely complex operation in the conditions prevailing in the 1950s. Doctors and nurses testified that there were cases where parents came to visit their children only to be told that their child had died. Without detailing the controversy of the state inquiry over the disappearance of a number of Yemenite children during this period – sparked by charges that the missing children may not have died, but had been given to childless couples for adoption – there is no question that these children, and the authority over their bodies imposed by the medical system, constituted a crucial juncture in the clash between immigrants and the state.[58]

Children, a core axis in Zionist perceptions of the rebirth of the Hebrew nation in its homeland, were viewed very differently by the new immigrants. It is easy to imagine that the high incidence of illness and death amplified parents' sense of need to protect their children in every way possible. Often the medical establishment was perceived as a supportive and beneficial agency, but reciprocal relations were very complex and the immigrants did not play a solely passive role in it. Thus, for instance, simplistic descriptions found in many historiographic accounts of the period that try to explain why children were transferred for care, such as 'at the beginning the parents were opposed to concentration of the sick children in separate camps, but in short order they were agreeable to such [a move]',[59] do not grapple with the complexity of the subject and neatly 'remove' the immigrants as key protagonists.

Conclusion

I do not intend to claim that in the 1950s there was wide-scale or 'organised' resistance to vaccinations among immigrants. It appears, in the end, that most immigrants were vaccinated, just as the overwhelming majority accepted other forms of medical intervention. But from reading primary and secondary source material relating to the years of mass immigration following the establishment of the State of Israel, one can reconstruct many cases of opposition that simply cannot be ignored. Vaccination, and particularly the vaccination of children who were the primary targets of the immunisation programme, demonstrates the way in which hegemonic machinery was used against immigrants and how the immigrants responded – and also opposed – these initiatives. Such resistance, or micro-resistance, happened not only in the realms of health but also in other spheres of life related to the settlement policy of the newcomers or education policy. The ethnic component, especially given the changing demography of the newcomers described above, needs to be carefully considered.[60]

Much has been written about the tie between Zionism and colonialism. In the Israeli context, the 'import', intellectually and physically, of Western medicine in the midst of an accelerated process of immigration from a large number of countries presents a unique case to scholars of health and immigration. This was particularly true with respect to the vaccination of children. Parents regarded vaccination as an act committed primarily 'against children', where their autonomy, authority and 'efficacy as parents' was usurped, leaving them in no position to object. In terms of understanding the relationship between the state, public health personnel and the population, it is important to acknowledge that over a long period the issue of vaccination was considered an important component of the colonial system. Despite good intentions, local populations frequently associated vaccination policy with a repressive regime. In addition, the very idea of introducing a disease-causing agent into the body of a healthy child could be perceived to be illogical and dangerous.

Immigrant populations were one of the key foci of public health agents in many countries. Arrival in a new country, a process that included medical examination in most cases, provided an opportunity to carry out mass vaccination of the immigrant population. The perception that immigrants as a group were less healthy than the local population served as the justification for an inflexible policy with regard to preventive medicine, including the adoption of mass vaccination. Thus opposition to vaccination should be situated within a broader debate tied to questions concerning the limits of state power in the private sphere – family life, religious belief and health – often accentuated by ethnic tensions as in the case of migrants. Indeed, oftentimes the attitudes of the vaccinators towards immigrant populations, the compulsive manner in which vaccinations were presented and the lack of sensitivity for the feelings of the target populations had a negative impact. More than once, such behaviour undermined trust and, as a consequence, made it hard for the vaccinators to convince people of the necessity of vaccination.

The 'marking' of the immigrant as 'other' by veteran populations – as a source of disease, crime and social ills that would 'pollute' host society – is a recurrent theme in various societies and periods. This 'otherness' has often been constructed through the lens of ethnic difference, thus amplifying questions of prejudice, racial discrimination and equality in access to health services. The context in which this appears varies across time and place. The State of Israel constitutes a unique test case of immigration and health issues. The tension between inclusion and exclusion tendencies towards the Jewish newcomers in the newly established Israeli state, between the Zionist ethos of the 'Ingathering of the Exiles' and the repulsion expressed in the fear of infections, created an ongoing tension expressed in responses to public health policies. The changing demographic composition of immigrants to Israel in the early 1950s amplified ethnicity with regard to issues such as the need for medical selection or the appropriate way of implementing public health practices, including vaccination. The racial/ethnic component of the historical debate surrounding the absorption of mass Jewish immigration to Israel is still very controversial and usually repressed.[61]

The body, once viewed almost as a 'natural' entity, assumed to be a neutral site, has become a productive research subject structured within various power networks and including an ethnic dimension.[62] Discourse on the Israeli body politic has also enjoyed more attention. The interaction between ethnic groups that sought to model a healthy and normal Israeli body opened up a broad new research field. The process of 'moulding the body' is complex and has included opposition in a host of intricate ways and by various groups. Recollections of 'micro-opposition' – in the daily events that took place in the absorption of new immigrants in the course of attempts to mould their body in the immigrant camps, in the schools, in the course of vaccinating and medical examinations – facilitate a socio-historical study capable of reconstructing the complex multi-faceted power relationships in play.[63] In most cases, opposition was local and the struggle took place in the realm of daily routine. Despite the challenges associated with this approach, it is important to attempt to reconstruct reciprocal relationships between the immigrants and the absorbing establishment: arenas of cooperation on one hand and tension and opposition on the other.

Even when many good intentions were involved in immigrants' absorption during the 1950s, the racial image of the newcomers suffering from social diseases, such as TB or ringworm, has continued to be an integral aspect of both medical and public discourses, an aspect that should not be ignored. The State of Israel has continued in more recent decades to absorb immigrants on an unprecedented scale. A host of questions asked in the 1950s, particularly vis-à-vis the reciprocal relationship between the state and those being absorbed, remain cogent to this day and deserve broader, interdisciplinary study. Moreover, if one adds to this the 'open wounds' that remain from the mass immigration of the 1950s, particularly in terms of ethnic struggle, such as that of missing Yemenite children and the mass ringworm irradiations, health and immigration certainly deserve a more prominent place in future historiographic investigations.[64] The controversy in the 1990s, when it became public knowledge that the Israeli blood bank was discarding blood donations from all Jewish Ethiopian immigrants,[65] again brought to the surface the relationship between public health policy, racial tensions and divergent perceptions of the body within the medical establishment and Israeli society and among immigrants.[66] It is not my intention to claim that 'history is repeating itself'; every case study related to health and immigration took place within its own cultural and social context and at a particular time, but there are common dilemmas and motifs that seem to indicate a phenomenology that demands analysis. Even when many good intentions were involved in immigrants' absorption, either during the 1950s or later, the racial image of newcomers suffering from social diseases such as TB or ringworm, diseases related to social backwardness, has continued to be an integral part of both medical and public discourse, and should not be ignored. Moreover, studies that examine waves of mass immigration to Israel need to take into account the perceptions of immigrants from different ethnic backgrounds, and how the divergent outlooks of immigrants impact on Israeli society. Questions relating to body perception, and the influence thereon of various hegemonic machineries, can contribute an often neglected dimension to these studies.

Notes

1. See Nadja Durbach, *Bodily Matters: The Anti-Vaccination Movement in England* (Durham, NC: Duke University Press, 2005); John Colgrove, *State of Immunity: The Politics of Vaccination in Twentieth-Century America* (Berkeley and Los Angeles, CA: University of California Press, 2006).

2. David Arnold, *Colonizing the Body: State Medicine and Epidemic Disease in Nineteenth Century India* (London, Berkley and Los Angeles: University of California Press, 1993).

3. On the civic potential of the history of anti-vaccination, see Robert D. Johnston, *The Radical Middle Class: Populist Democracy and the Question of Capitalism in Progressive Era, Portland, Oregon* (Princeton, NJ: Princeton University Press, 2003), 177–220.

4. See Warwick Anderson, *The Cultivation of Whiteness: Science, Health and Racial Destiny in Australia* (New York: Basic Books, 2003). On the concept of 'Glocalisation', see Roland Robertson, 'Glocalisation: Time-Space and Homogeneity-Heterogeneity', in Mike Featherstone, Scott Lash and Roland Robertson (eds.), *Global Modernities* (London: Sage, 1995).

5. *Ma'abarot* (transit camps) were used as 'temporary' housing communities of tin and wooden shacks until permanent housing could be built. In practice, they remained in place until the early 1960s. Other immigrant camps were *machanot maavar* (processing camps) and *machanot olim* (immigrant camps) where newcomers were housed in tents.

6. At the end of 1948, the number of available beds for TB patients in Israel stood at 300, less than 10 per cent of those needed. Labor Archives, Levon Institute IV-137-3-243, letter from Dr Meir to the Minister of Health and Immigration, 28 April 1949.

7. On the concept of the 'melting pot', see Moshe Lissak, *The Mass Immigration in the Fifties: The Failure of the Melting Pot Policy* (Jerusalem: Bialik Institute, 1999) (in Hebrew); Tzvi Tzameret, *The Melting Pot: The Frumkin Commission on Education of Immigrant Children (1950)* (Ben-Gurion Research Center: Ben Gurion University Press, 1993) (in Hebrew).

8. See Nadav Davidovitch and Shifra Shvarts, 'Health and Hegemony: Preventive Medicine, Immigration and the Israeli Melting Pot', *Israel Studies*, 9 (2004), 150–79.

9. See also Rina Peled, *'The New Man' of the Zionist Revolution: Hashomer Hatzair and its European Roots* (Tel Aviv: Am Oved, 2002) (in Hebrew).

10. On eugenics and Zionist ideology, see Raphael Falk, 'Zionism and the Biology of the Jews', *Science in Context*, 11 (1998), 587–607; Raphael Falk, 'Eugenics and the Jews', in Alison Bashford and Philippa Levine (eds), *The Oxford Handbook of the History of Eugenics* (Oxford: Oxford University Press, 2010), 462–76; John Efron, *Defenders of the Race: Jewish Doctors and Race Science in Fin-de-Siècle Europe* (New Haven, NJ: Yale University Press, 1993).

11. See Anderson, *The Cultivation of Whiteness*.

12. See Dafna Hirsch, '"We Are Here to Bring the West, Not Only to Ourselves": Zionist Occidentalism and the Discourse of Hygiene in Mandate Palestine', *International Journal of Middle East Studies*, 41 (2009), 577–94.

13. Quotation from David, speaking before the Knesset and cited in Dvora Hacohen, 'The Law of Return – Its Content and the Debate Surrounding It', in Anita Shapira (ed.), *Independence: The First 50 Years* (Jerusalem: Zalman Shazar Center for the History of Israel, 1998), 57 (in Hebrew).

14. Law of Return (1950), clause 2 (a) (2).

15. The issue of the selection of immigrants on medical grounds had surfaced in the 1920s and 1930s but not as open public debate as was the case in the 1950s. See Shifra Shvarts, Nadav Davidovitch, Avishay Goldberg and Rhona

Seidelman, 'Medical Selection and the Debate over Mass Immigration in the New State of Israel', *Canadian Bulletin of Medical History*, 22 (2005), 5–34. The changing demography of the immigrants meant that when medical selection was implemented in 1951 it impacted mainly on Jews arriving from Arab countries. The policy of individual medical selection was abolished in 1954.

16. Dr Yosef Meir, 'Medical Plan for Absorbing New Immigrants, 24.12.1943', in Dvora Hacohen (ed.), *From Fantasy to Reality: Ben-Gurion's Plan for Mass Immigration 1942–1945* (Tel Aviv: Ministry of Defense, 1995), 264–5 (in Hebrew). This document addressed the issue of vaccinations of immigrants at absorption stations: 'In every station there should be a setup of doctors, nurses, sanitation personnel and social workers to oversee and carry out disinfection, treatment and vaccination against typhus, smallpox, diphtheria and so forth'.

17. See Theodore Grushka, *Health Services in Israel: A Ten Year Survey, 1948–1958* (Jerusalem: Ministry of Health, 1959), 113–21.

18. See Shifra Shvarts, *Kupat Holim, the Histadruth, and the Government: The Formative Years of the Health System in Israel, 1947–1960* (Ben-Gurion Research Center, Ben Gurion University Press, 1999) (in Hebrew).

19. See the description of this phenomenon in Labor and Pioneer Archive, Levon Institute, IV-243 Folder 3–6, letter from Dr Y. Shertok to the Minister of Health, 11 May 1949.

20. On the establishment of the *Shaar Ha'aliya* camp and its public health function, see Rhona D. Seidelman, S. Ilan Troen and Shifra Shvarts, '"Healing" the Bodies and Souls of Immigrant Children: The Ringworm and Trachoma Institute, Sha'ar ha-Aliyah, 1952–1960', *Journal of Israeli History: Politics, Society, Culture*, 29:2 (2010), 191–211; Rhona D. Seidelman, 'Conflicts of Quarantine: The Case of Jewish Immigrants to the Jewish State, *American Journal of Public Health*, 102 (2012), 243–52.

21. See Grushka, *Health Services in Israel*, 115. See also Seidelman, Troen and Shvarts, '"Healing" the Bodies and Souls of Immigrant Children'.

22. Abraham Sternberg, *As a People is Absorbed* (Tel Aviv: Hakibbutz Hameuchad, 1973), 124–5 (in Hebrew).

23. See Grushka, *Health Services in Israel*, 116.

24. See Nadav Davidovitch and Avital Margalit, 'Public Health, Racial Tensions, and Body Politic: Mass Ringworm Irradiation in Israel', *Journal of Law, Medicine & Ethics*, 36 (2008), 522–9.

25. See Yehuda Weissburger, *Shaar Haaliyah – Mass Aliyah Diary, 1947–1957* (Jerusalem: World Zionist Organization, 1986), 71.

26. See, for instance, Labor and Pioneer Archive, Levon Institute, IV 104–81, folder #17, Memorandum on a visit to Eden, Ministry of Health, 20 September 1949.

27. In Tunisia, for instance, only smallpox vaccination was compulsory by law while vaccination against typhoid or diphtheria required parental permission. In practice, due to low levels of compliance, typhoid shots were not given. See Raphael Greenberg, 'Report on Medical and Social Problems in Tunisia' presented at the *American Joint Distribution Committee Medical Conference*, 28th June –1st July, 1954, 18, AJDC Archives, NY, Box 123, Africa – Tunisia – Medical, 1948–1960.

28. Public Health Ordinance, 1940. See also State Archives, 5084/13/2, letters from Dr P. Yekutiel to the Health Bureau, smallpox vaccination 26 February 1956, and from R. Klezmer to the Head of Implementation Service, Requirement for Vaccination of Infants against Smallpox, 10 May 1959.
29. In the mid-1950s vaccination against typhoid was given with diphtheria. See also *Immigration Regulations*, State of Israel, Ministry of Health, Jerusalem, 1955, Chapter VI Vaccinations, 5. Subsequently, vaccination against diphtheria was integrated into the general vaccination programme with tetanus and pertussis, known as DTP vaccination. In addition, immigrants arriving from 'affected' countries where other specific diseases such as cholera and yellow fever were prevalent, required vaccination against these diseases as well. See *Immigration Regulations*, Appendix 2, 1–3.
30. See Labor and Pioneer Archives, Levon Institute, IV 104–669, folder #17, *New Immigrants – Levy among Immigrants in the Ma'abarot and in Working Villages and the New Immigrants's Rights*, General Federation of Labor, Sick Fund Central Committee – Finance Department, Tel Aviv, 28 January 1951.
31. See State Archive, 5084/13/2-3, Dr Y. Shapira to Dr Yekutiel, *Smallpox Vaccination for New Immigrants*, 24 January 1956 (in Hebrew).
32. See Hadassa Heinreich, 'The Vaccinations Customary in *Eretz-Israel*', *Dapim Refuim*, 9 (1950), 167–75 (in Hebrew).
33. Essentially smallpox vaccination techniques have not changed since that period. This issue re-emerged when the Israeli Ministry of Health decided to vaccinate against smallpox the 'first responders' in a case of a bioterrorist attack in 2003.
34. Interview with B.S., a doctor who practised at the time and worked in immigrant camps in the 1950s. The physician said that at times they vaccinated girls under the nipple for cosmetic reasons, to hide the scar. No sign of this as a 'method' was found in contemporary medical textbooks, but the medical literature of the period mentioned the possibility of vaccinating girls under the knee, should the mother demand this for cosmetic reasons. The authors usually do not recommend these methods due to the pain and sensitivity involved. See *Merck Manual of Diagnosis and Therapy* (Rahway, NJ,: Merck, 1950), 680–1.
35. IDF Archives and the Ministry of Defense, 110-648/53, A. Kleinberg, *Instruction for Vaccination against Smallpox, 1950* (in Hebrew).
36. See, for instance, State Archives, 5084/13/2-3, the letter classified 'secret – very urgent' from Dr P. Yekutiel, Preventive Measures against Smallpox, 21 September 1953 (in Hebrew).
37. For information on the degree of public response to calls for vaccination against typhus, see 'Instructions for Compulsory Vaccination against Typhoid: Difficulties in Sanitation Activities throughout the Country', *Haaretz*, 9 April 1952: 'Last year's attempt to reach the public to vaccinate via propaganda failed in essence. Last year only 200 souls received the required vaccination, at least 2/3 of them schoolchildren and youth, that is the response among adults was almost nil'. The article emphasised that the primary danger was tied to 'mass immigration and our economic straits'.
38. On the controversy over typhoid vaccination and hygiene, see *Hygiene and Health*, 1 (1940), 8–9.

39. See State Archives, 4520-2/5/4, letter from Dr P. Yekutiel to the Director General, Ministry of Health.
40. See Herman Lichtenstein, 'Results of Mass Reontgenography among Immigrants into Israel', *American Review of Tuberculosis*, 69 (1954), 837–40.
41. See Labor Archives, Levon Institute, IV-137-3-243, letter from Dr Yosef Meir to the Minister of Health and Immigration, 28 April 1949 (in Hebrew).
42. *Mass BCG Vaccination in Israel, 1949–1950: Campaign Carried Out Under the Joint Auspices of the Ministry of Health of Israel, Danish Red Cross, Norwegian Relief for Europe, Swedish Red Cross and United International Children's Emergency Fund*, published by the International Tuberculosis Campaign, Neuilly-sur-Seine, France: UNICEF EMRO, 1953. See Roberta Bivins' chapter in this volume, for campaigns against TB in Britain.
43. See, for instance, Grushka, *Health Services in Israel*.
44. See Dvora Hacohen, *Immigrants in Turmoil: Mass Immigration to Israel and Its Repercussions in the 1950s and After* (Syracuse, NY: Syracuse University Press, 2003).
45. On the state of infants in the transit camps, see Sternberg, *As a People is Absorbed*, 38–68 (in Hebrew), see also State Archives, 4265/188/7, letter from Dr Towstein to Dr Lotan on the health status of infants in the transit camps, 16 July 1952 (in Hebrew). For conditions in the *Ma'abarot* from the perspective of doctors who served there, see also State Archives, 4265/188/7, Dr Emanuel Margalit and Dr Betzalel Pinchas, Reflections on Medical Service in the Ma'abarot, n.d (in Hebrew).
46. State Archives, 4250-2/6/4, meeting of the Coordination Institute meetings, Dr Chaim Sheba 11 July 1951 (in Hebrew).
47. *Haaretz*, 5 July 1950 (in Hebrew).
48. Sternberg, *As a People is Absorbed*, 98–9 (in Hebrew).
49. Hannah Rosental Munk, *The Response of Public Health Nursing to Mass Immigration in Israel 1948–1958* (unpublished PhD thesis, Columbia University, 1979), 232.
50. See Meira Weiss, 'The Immigrating Body and the Body Politic: The "Yemenite Children Affair" and Body Commodification in Israel', *Body & Society*, 7 (2001), 93–109.
51. The second juncture when a significant portion of the population receives vaccinations is upon induction into the Israel Defense Forces (IDF).
52. See Shifra Shvarts, 'The Development of Mother and Infant Welfare Centers in Israel, 1854–1954', *Journal of the History of Medicine and Allied Sciences*, 55 (2000), 398–425.
53. Ben-Gurion Heritage Archive, Diary of a Mentee in the Zisor Maabara, ID Number 180469, 30–31 January1951 (in Hebrew).
54. See, for example, Emily K. Abel and Nancy Reifel, 'Interactions between Public Health Nurses and Clients on American Indian Reservations during the 1930s', *Social History of Medicine*, 9 (1996), 89–108.
55. See Phyllis Palgi, 'How It All Began ... A Personal Saga', *Practicing Anthropology*, 15 (1993), 5–8.
56. See Lenore Manderson and Pascale Allotey, 'Storytelling, Marginality, and Community in Australia: How Immigrants Position Their Difference in Health Care Setting', *Medical Anthropology*, 22 (2003), 1–21.
57. See Weiss, 'Immigrating Body and the Body Politic', 98

58. The health system was not the only one representing this 'struggle'. Other hegemonic systems, such as the educational system, the army and other social institutions, constituted additional contested realms.

59. Hacohen, *Immigrants in Turmoil*, 282. An example of a similar 'explanation' can be found in an article that appeared in the newspaper *Davar* that describes the medical care of immigrants. Among other things, the article states: 'One can cite with satisfaction that women from Yemen who at the beginning did not want to give their children to the infant house, began to fathom the great blessing of these institutions and now they give their children to the infant house willingly and happily'. A. Ch. Alchananai, Medical Assistance to Immigrants: the How, *Davar* newspaper, 25 August 1950, 8 (in Hebrew).

60. On micro-resistance related to the settlement policy of the newcomers in the 1950s, see Adriana Kemp, *Nedidat Amim oh ha'Be'era ha'Gdola: Shlita Medinatit ve'Hitnagdut ba'Sfar ha'Tsraeli*, in Hannan Hever, Yehouda Shenhav, Pnina Motzafi-Haller (eds.), *Mizrahim in Israel: A Critical Observation Into Israel's Ethnicity* (Tel Aviv: Van Leer Jerusalem Institute and Hakibbutz Hameuchad, 2002), 36–65 (in Hebrew).

61. See Yehuda Shenhav, *The Arab Jews: Nationalism, Religion and Ethnicity* (Stanford, CA: Stanford University Press, 2006); Sami Shalom Chetrit, *Mizrahi Struggle in Israel, 1948–2003* (Tel Aviv: Am Oved, 2004).

62. There is already a rich and productive tradition of 'body politics' writing in health research. See, for example, M. Lock, 'Cultivating the Body: Anthropology and the Epistemologies of Bodily Practices and Knowledge', *Annual Review of Anthropology*, 22 (1993), 133–55.

63. See also Kemp, 'Nedidat Amim'.

64. On the racial/ethnic aspect of the ringworm affair, see Davidovitch and Margalit, 'Public Health, Racial Tensions, and Body Politic'.

65. Secretly discarding blood donations was based on epidemiological criteria, namely the prevalence of AIDS in their country of origin, but the policy was kept confidential in order not to 'taint' the ethnic group's image.

66. Don F. Seeman, *One People, One Blood: Ethiopian-Israelis and the Return to Judaism* (New Brunswick, NJ: Rutgers University Press, 2010).

8
From the Cycle of Deprivation to Troubled Families: Ethnicity and the Underclass Concept

John Welshman

Introduction

Research on the history of ethnicity and health has begun to point to the atypicality of the policy response of the United Kingdom, especially compared with the experiences of other countries such as Australia and the United States. In the specific case of tuberculosis (TB) screening of migrants between 1950 and 1965, for example, civil servants at the Ministry of Health managed to resist and subvert pressure in favour of compulsory medical examinations at the ports of entry and set up a different type of screening system at the local level. It was not so much the theme of compulsion, often emphasised in the historiography of both TB and public health, but those of localism and pragmatism that proved decisive in the adopted UK policy stance of screening after arrival.[1]

This chapter seeks to move beyond policy and to further explore the presence – and absence – of ethnicity in broader and changing discourses around the notion of an 'underclass' in Britain from the 1970s. While the primary focus of the chapter is the United Kingdom, the United States offers an illuminating contrast, and so some limited reference is made to similar debates there in the 1960s and the 1980s. It is generally accepted now, by historians of public policy, that there have been a series of labels, stretching from the 'social residuum' of the 1880s, through the 'social problem group' of the 1920s, the 'problem family' of the 1950s, the 'cycle of deprivation' of the 1970s and the underclass of the 1980s to the contemporary emphasis on 'troubled families', that together form a series of conceptual stepping stones through which the concept of the underclass has been successively invented and reinvented in modern Britain.[2] Indeed this appears to be a national discourse peculiar to Britain and the United States, part perhaps as a legacy of the 'Anglo-Saxon' empirical approach to poverty typified by researchers such as Seebohm Rowntree, with his social surveys of York, and Peter Townsend, whose work was both empirical and conceptual.[3] While the term *'underklass'* originated in Sweden in the late nineteenth century, and while social exclusion became popular in France in the early 1970s, it is much more difficult to find an equivalent discourse in other European countries. In a similar fashion, research has sought to explain why there was no underclass in Australia in the 1980s; its absence has been explained by the existence of a high wage economy, low unemployment, and the (claimed) relatively successful incorporation of migrant groups into Australian society.[4]

There are a number of important strands to the concept and to the way it has operated in both the United Kingdom and the United States. First is the way that concepts have been used to signify and denote the behavioural inadequacies of the poor. Second is their use to denote the ways in which wider structural processes have contributed to a situation in which groups with poor access to education or skills risk being left behind. Third, and perhaps most importantly, there is the recurring belief in inter-generational continuities. Fourth, we can trace the belief that the underclass exists separately from the working class. And fifth, what is striking is the combination of rhetorical symbolism (where the concept operates at a metaphorical level for middle-class fears and anxieties) and empirical complexity (where there are real changes related to family size and formation, poverty, housing, employment levels and so on).[5] The way that the concept has been defined in different periods has said as much about broader trends in the economy and the labour market, the role of women and the emphasis placed on the nuclear family, migration, urbanisation, and ideas about behaviour and agency, as about the underclass itself. At various times issues such as joblessness, household squalor, mental health (and learning disability), long-term poverty, illegitimacy and crime have all been drawn into underclass stereotypes. But there are also continuities, in terms of the alleged physical and mental characteristics of the poor, the stress placed on inter-generational continuities, the focus on behavioural inadequacies, the emphasis on the costs to the state and the desire to quantify the size of the 'problem' (for example, the underclass have often been perceived as being the bottom 10 per cent of society).

The terms appear to follow a trajectory of initial popularity, currency, the acquiring of a pejorative connotation, a falling out of favour, and then a process of replacement by another (apparently less pejorative) term. In fact, there have been few periods since the 1880s when at least some variant of the underclass concept has not been available to observers and commentators. How and when such concepts emerged is more complex. Often it was during a period of economic slump and unemployment, such as during the 1920s. Interest in the social residuum fell away during the First World War, as those apparently 'unemployable' (including people with disabilities) were drawn into the workforce. We can also date very precisely the birth of the problem family to the *Our Towns* survey of evacuation, published in 1943 by the Women's Group on Public Welfare, which both argued for the expansion of health services and blamed mothers for the poverty of their families.[6] Possibly this says something about the nature of the differences between the First and Second World Wars – the latter being much more a 'people's war' on the Home Front. More generally, the history of the discourse of the underclass can provide a window through which to observe the processes of policy transfer, and specifically the transfer of ideas, if not of policies, from the United States to the United Kingdom.[7] Nevertheless, while in some periods the various concepts have had a limited impact on policy making, both at the local authority and national government level, in others the discourse has not had any clear impact on policy.

It is perhaps not particularly surprising that ethnicity played relatively little part in earlier underclass concepts. The comments of the social investigator, Charles Booth, on the Jewish population of Tower Hamlets in London are helpful. In 1887, for example, he noted that the 'English Jews' were mainly well-to-do; the Dutch were 'not very poor'; and the Germans were 'a striving, hardworking set of people, and, on the whole, prosper'. While the 'Russian Poles' were employed as tailors and arrived destitute, they rarely applied to the Poor Law Guardians for assistance, relying on Jewish charities, notably the Jewish Board of Guardians, which catered specifically for the Jewish community. Overall, Booth wrote, these people had the characteristics of their 'race', in that they were 'laborious', frugal, able to live on next to nothing and hard working.[8] Crucially, given the importance of notions of dependency to the underclass concept over time, the Jewish population was perceived as producing more than it consumed. Certainly there were powerful negative stereotypes of Irish migrants from the mid-nineteenth century, and general prejudice, but rather surprisingly perhaps, the Irish as such were not drawn into the narrower notion of the social residuum in the 1880s nor the debate about unemployables in the early 1900s.[9]

Similarly, the social problem group of the interwar period seemed to be free of ethnic prejudices, even if its strongly eugenic emphases focused on mental deficiency and the social 'problems' of unemployability, dependency, vagrancy and prostitution. In this respect, the debate in Britain was markedly different to that in Nazi Germany. Moreover, while the focus on large families in the problem family debate of the 1940s and 1950s might be seen as a proxy for ethnic minorities (and Irish Catholics), again this seemed to be perceived as being predominantly white families. The survey of problem families sponsored by the Eugenics Society in the late 1940s, for instance, focused on areas that included Hertfordshire, Luton and Rotherham, not necessarily locales with sizeable ethnic populations. Pat Starkey has argued of Bristol that there were changes in the type of family that the label was used to describe, with families from ethnic minority groups being increasingly identified in this way by the mid-1960s.[10] Nevertheless, research on the Midlands city of Leicester (one with a large ethnic minority population) provided little evidence to support Starkey's contention.[11]

This chapter seeks to contribute to the history of health and ethnicity by considering the emphasis given, or not given, to ethnicity in the changing 'underclass' discourse in the period since the early 1970s. It surveys the 1974–1982 Research Programme on Transmitted Deprivation, funded by the Department of Health and Social Security, and organised by the Social Science Research Council, that followed Sir Keith Joseph's famous 'cycle of deprivation' speech of June 1972. While the speech and the Research Programme have attracted attention in recent years, not least because of the striking continuities between Joseph's ideas and the policy emphases of the Labour and Coalition Governments since 1997, the part played by ethnicity has so far not been given the same attention.[12] The chapter looks at debates about an underclass in the 1980s in the United States and the United Kingdom. It then juxtaposes the 1970s debate with research on social exclusion in the United Kingdom since the mid-1990s. The argument is that, for a variety of reasons, and despite the empirical evidence of the very real disadvantage experienced by ethnic minority groups, ethnicity has generally been underplayed in this discourse (Figure 8.1). In this way, the presence and omission of ethnicity in the underclass discourse in the United Kingdom over the past 40 years offers broader insights into both the empirical reality of the ethnic experience and the response to it, which was largely conditioned by political and ideological factors.

The cycle of deprivation debates of the 1970s

Our starting point is the debate about the 'cycle of deprivation' or 'transmitted deprivation' in the United Kingdom in the 1970s. In terms

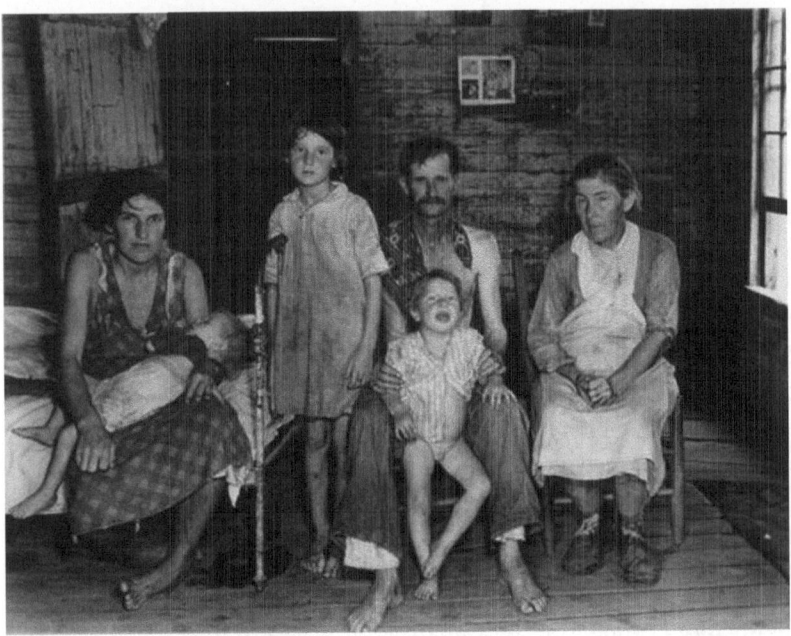

Figure 8.1 'Bud Fields and His Family at Home', Photographed By Walker Evans (1903–1975), Alabama, 1936. Courtesy of the Library of Congress, Prints & Photographs, LC96516419.

of the cycle of deprivation, the key speech by Sir Keith Joseph, then Conservative Secretary of State for Health and Social Services, was given on 29 June 1972. It was in the second half that Joseph developed his main theme, asking why it was 'that, in spite of long periods of full employment and relative prosperity and the improvement in community services since the Second World War, deprivation and problems of maladjustment so conspicuously' persisted.[13] By deprivation, Joseph meant 'those circumstances which prevent people developing to nearer their potential – physically, emotionally and intellectually – than many do now'.[14] He acknowledged that deprivation took many forms and had complex causes, including those that were economic, personal and related to patterns of child rearing. But, he continued, 'perhaps there is at work here a process, apparent in many situations but imperfectly understood, by which problems reproduce themselves from generation to generation'.[15] There was not a single process. But it seemed that in a proportion of cases, the problems of one generation were repeated in the next. Part of Joseph's speech called for more research, and also recognised that the 'cycle' was poorly understood. Though Joseph did acknowledge that poverty played a role in the causation of deprivation, his remedies were noticeably more limited. Apart from playgroups and services for the under-fives, they focused on family planning, support for parents and attention to the needs of children.[16] Perhaps not surprisingly, Joseph did not mention ethnicity.

The speech was reported in the main newspapers, but met with a fairly muted response. Nevertheless it did lead to a Research Programme on Transmitted Deprivation, which ran for eight years, 1974–1982, and cost £0.75m at 1970s' values. A literature review, completed in 1973 and published in 1976 by Michael Rutter from the Institute of Psychiatry in London and his Research Assistant, Nicola Madge, examined evidence that might support the 'cycle of transmitted deprivation' and considered what it was that created these alleged continuities between generations. Rutter and Madge decided that they preferred the term 'disadvantage' to the original 'deprivation'; they substituted the plural 'cycles' for the singular 'cycle' and they dropped the phrase 'transmitted'.[17] These changes would have an important bearing on the Research Programme as a whole. Their survey covered physical health; parenting and child care behaviour; psychiatric disorders; crime and delinquency; ability, attainment and education; employment; housing and discrimination in housing; and race relations, and they included a lengthy chapter on ethnic minorities in Britain.[18] They argued that 'immigrant populations present an opportunity to determine how various social, psychological and economic forces lead individuals into, or protect them from, disadvantage and deprivation'.[19] Some processes were related to the fact of immigration, some to skin colour and associated prejudice and discrimination, and some to circumstances shared with other groups in the population. The survey was generally cautious about research carried out in the United States, pointing to the differences with Britain in terms of history and 'social climate', and in the lifestyles of black families; for instance, the term 'ghetto' seemed not to be applicable to Britain. The possibility of a genetic component to the difference in average IQ between blacks and whites was not excluded, but equally there was no convincing evidence in favour of the proposition.[20] Some of the difference, however, was accountable for in terms of environmental influences.

The Joint Working Party on Transmitted Deprivation that was set up to organise the research noted in its First Report of August 1974 that research was needed on 'colour'. Ethnicity was thus clearly seen in terms of the non-White population. Skin colour was a strong indicator of deprivation of various kinds, but research was needed into the circumstances in which it existed, the reasons for the connection and the extent to which these connections were involved in the transmission of problems across generations.[21] Nevertheless, the Joint Working Party did not commission much, if any, research on ethnicity, and in its Third Report of November 1977 said it would simply commission a literature review on 'the influence of ethnic origin'.[22] The Joint Working Party did not, therefore, appear overly interested in ethnicity in relation to transmitted deprivation, and the literature review was a belated attempt to remedy this.

The review was carried out by Alan Little and Diana Robbins, then respectively Lewisham Professor of Social Administration and Research Officer at Goldsmiths' College, University of London, and it was published by the Commission for Racial Equality. Little and Robbins followed the lead given by Rutter and Madge in their scepticism about the term 'transmitted deprivation', and also about the differences between Britain and the United States. Little and Robbins argued that a study of transmitted advantage would be relatively uncontroversial, whereas a study of transmitted deprivation was immediately contentious for political and ideological reasons. Racial or ethnic discrimination was, in the view of the authors, compounded by sex and sectarian discrimination. The kind of structural transmission that the authors were concerned with was the unfavourable attitudes of the majority to identifiable groups, giving rise to negative stereotypes and discriminatory behaviour that limited the opportunities of members of those groups. Nevertheless, the study did not reach firm conclusions about policies that would have to be implemented to break cycles of discrimination-induced disadvantage. They did, however, note that a research project undertaken by the social survey organisation, Political and Economic Planning between 1972 and 1975, had shown the disadvantages of ethnic minority groups with regard to employment, housing and racial disadvantage.[23]

Little and Robbins thus took a stance that was primarily structural. Elsewhere, in an overview of this work, Little and Robbins noted that there was evidence of continuing inequalities over time, but also of a widening gap between the life chances of adult and adolescent black people, and between the black and white sections of society. The negative evaluation by the white majority of genetically transmitted skin colour, as well as of socially transmitted customs, skills, religious beliefs, arts and language, was paralleled and compounded by sex and culture-based sectarian discrimination. It was this negative evaluation that provided an extra handicap for minorities already caught up in insecure, low-paid or low-status employment, bad housing, the poorest education and the environmental deprivation of the inner city. In particular, Little and Robins compared the situation of ethnic minorities to that of women, and to Catholics in Northern Ireland (whose discrimination in such areas as housing compared to the Protestant majority had inspired much of the civil rights activity of the late 1960s). Much of this was in the context of the 1975 Sex Discrimination Act, the 1976 Race Relations Act, and the recognised need for affirmative action or area-based positive discrimination.[24]

Nevertheless, when they reviewed the Transmitted Deprivation Research Programme as a whole ten years on, in 1982, Muriel Brown and Nicola Madge, then based at the London School of Economics (LSE), only referred occasionally to ethnicity. This was in relation to cultural values, discrimination, educational achievement, employment, family profiles, health, housing and income. In each of these, the authors were at pains to refute a behavioural interpretation and to emphasise instead what they saw as structural factors.[25] This was perhaps not especially surprising, given that ethnicity had not been a major component of the Research Programme, and the literature review by Little and Robbins had been almost an afterthought. Reflecting on his Chairpersonship of the Supplementary Benefits Commission, David Donnison recalled that, with hindsight, the early 1970s stood out at the end of a period that had begun when he was a child in the 1930s. The debate about social security policies was for many years shaped by the assumptions of that time. People who worked in the 'Titmuss school' of social scientists, inspired by Richard Titmuss, Professor of Social Administration at the London School of Economics, found that they had joined a much larger group of progressive social democrats who shared similar concerns and assumptions. These were that the growth of the economy and the population would continue, the harsher effects of inequalities in incomes would gradually be softened by a 'social wage' and a growing burden of progressive taxes, and middle England would eventually support equalising social policies and programmes of this kind. It was also assumed that government and its social services were the natural vehicles of progress, that their social policies could redistribute the fruits of economic growth, manage the human effects and compensate the disadvantaged, and that governments which allowed a return to the high unemployment, social conflicts and means tests of the 1930s would not survive.[26] In fact, although Donnison dates this shift to the early 1970s, the outlook of many social scientists continued to be shaped by the Titmuss paradigm into the 1980s and beyond.

The culture of poverty controversy in the United States in the 1960s

One reason for this tentative approach to ethnicity was the relative size of the ethnic minority community. The population of Great Britain of New Commonwealth and Pakistani origin (that is, the non-White population) was estimated in 1976 at 1.6m of a total population of 54.4m (only 3 per cent of the total).[27] But another factor was the response of British social scientists to related debates in the United States. Fieldwork in Puerto Rico gave Oscar Lewis, Professor of Anthropology at the University of Illinois, the chance to test out his theory of the culture of poverty, and the classic account appeared in the introduction to *La Vida* (1966). In this work, Lewis compared 100 low-income Puerto Rican families from four slums in Greater San Juan with their relatives in New York. He wrote that as an anthropologist he had tried to understand poverty as 'a culture or, more accurately, as a subculture with its own structure and rationale, as a way of life which is passed down from generation to generation along family lines'.[28] The culture of poverty was not just a matter of economic deprivation, but had a positive connotation. It had advantages for the poor, and indeed it was arguable that without it, they would be unable to carry on. Thus the culture of poverty was 'both an adaptation and a reaction of the poor to their marginal position in a class-stratified, highly individuated, capitalistic society'.[29]

Lewis argued that one characteristic of adults as opposed to children was the way that the poor did not participate in, nor were integrated by, the major institutions of the larger society. People with the culture of poverty, it was alleged, did not belong to trade unions, were not members of political parties, were not participants in the welfare system, and did not make use of banks. It was this 'low level of organisation' that gave the culture of poverty its marginal quality in a highly complex and organised society. Even so, Lewis was quick to point out that the culture of poverty was not just an adaptation. Once established, it tended to perpetuate itself through the generations, because of its effect on children. By the age of six or seven, argued Lewis, children 'have usually absorbed the basic values and attitudes of their subculture and are not psychologically geared to take full advantage of changing conditions or increased opportunities which may occur in their lifetime'.[30] Overall, Lewis argued that improved economic opportunities were not the whole answer. It was easier to eliminate poverty than the culture of poverty. His emphasis clearly influenced the 'War on Poverty', begun in the Kennedy administration and continued in the Johnson administration, through its emphasis on community programmes rather than large-scale economic and social change.

Earlier writing on black families also exerted an important influence on debates in American social policy in the 1960s. Franklin E. Frazier's *The Negro Family in the United States* (1939), in particular, went through numerous editions and became known to successive generations of social scientists.[31] But its most direct link with policy came with the Moynihan Report on *The Negro Family*, published in 1965. Its author, the US Republican Senator Daniel Patrick Moynihan, wrote that 'Negro social structure, in particular the Negro family, battered and harassed by discrimination, injustice, and uprooting, is in the deepest trouble'.[32] A quarter of urban black marriages were dissolved, one in four black births was illegitimate, and women headed a quarter of black families. Overall, Moynihan claimed that the breakdown in the black family had led to a startling increase in welfare dependency. Noting that 14 per cent of black children, compared to 2 per cent of white children, were in receipt of Aid for Families with Dependent Children (AFDC), Moynihan argued that the steady expansion of this welfare programme charted the steady disintegration of Negro family structure in the previous generation. At the centre of the 'tangle of pathology' was the weakness of family structure.[33] It has been suggested that the publication of the Moynihan Report led to an 'intellectual void', in which liberal scholars were deterred from discussing issues of individual behaviour, a vacuum that was filled by neo-Right commentators in the 1970s and 1980s.[34] Certainly British social scientists such as Peter Townsend remained sceptical about Oscar Lewis's culture of poverty, claiming that it was unlikely to be relevant to Britain. Michael Rutter and Nicola Madge concluded that 'the culture of poverty concept is inadequate for an analysis of British society'.[35]

The underclass debate in the United Kingdom in the 1970s and 1980s

It is, nonetheless, correct that when the term 'underclass' was used in discussions of British society in the 1970s, it was usually in relation to the position of ethnic migrant workers. In 1973, the sociologist Anthony Giddens argued that where ethnicity served as a 'disqualifying' factor in the labour market, and where ethnic groups were concentrated in the poorest paid jobs, or were unemployed or semi-employed, it was possible to talk about an underclass. Drawing on research carried out in Detroit,

Giddens argued that these 'distributive groupings' were formed by neighbourhood clustering and by certain combinations of status group formation. Their form varied according to differences in the size and density of urban areas, and in the social and political structures of capitalist societies. Giddens suggested that the existence of a large underclass cut across any clear-cut distinction between middle- and working-class neighbourhoods. He accepted that the size and demographic composition of the United States made it something of a special case. Nevertheless, Giddens maintained that in many European countries too, the lack of an indigenous ethnic minority led to a transient underclass being imported from outside. Similar developments were visible, Giddens contended, in Britain and France. Composed of recent migrants in urban and industrial areas, the underclass formed the basis for a pool of highly 'disposable' labour. Its members had few educational qualifications and were unskilled manual workers. Moreover, if migrant workers carried out certain jobs, the existence of this underclass made it easier to separate out jobs that would be more acceptable to the working class. Like his counterparts in the United States, Giddens was interested in whether this underclass was of potential political significance. He argued that the underclass could be viewed as a force for revolutionary change or as reinforcing conservative attitudes. Giddens predicted that 'hostile outbursts' were possible, because the underclass was unable to exercise the kind of citizenship rights enjoyed by everyone else.[36] Overall, despite its radical potential, he concluded that social unrest was unlikely.

Other commentators on Britain in this period, though, were more hesitant about using the term underclass to describe the position occupied by ethnic minority groups. In 1975, the sociologists John Westergaard and Henrietta Resler pointed out that ethnic minorities in Britain were not concentrated uniformly at the bottom of the social order. Although they undoubtedly faced serious obstacles in the labour market, as indeed in society in general, they did not constitute an underclass. Descriptions of migrant labour as an underclass, they suggested, had more application to continental countries, such as Germany and Switzerland, that had recruited large numbers of foreign workers into poorly paid jobs.[37] When the term 'underclass' was used in Britain it was deployed in a rather different way, with a more positive connotation. In their Birmingham case study, the sociologists John Rex and Sally Tomlinson argued that there was much evidence that migrants were discriminated against and also stigmatised in the way that the welfare state operated. They acknowledged there was some tendency for the black community in Britain to operate as a separate class or underclass, but resisted the idea that its members were an inert mass with a ghetto mentality or a culture of poverty. Rather, they argued that ethnic minorities organised and acted in their own underclass interests. Two factors pulled migrant workers away from a quasi-Marxist 'underclass for itself' concept; these were the affiliations that migrant workers had with the mainstream working class, and the influence of their homelands. Rex and Tomlinson concluded that the car industry in Birmingham provided unstable but highly unionised conditions, with good wages for workers. Other factories and foundries had few unions and provided work that was poorly paid but secure. They claimed that distinctions of this kind in the labour market were a 'necessary but not sufficient condition' of the emergence of an underclass.[38]

In fact, the British underclass debates of the 1980s focused much more on long-term unemployment in the white working-class population. Even the American commentator and polemicist Charles Murray noted on his visit to the United Kingdom that, unlike the United States, the role of ethnicity was much less significant in explaining out-of-wedlock births. Out-of-wedlock births were higher in the black community, but he conceded that this represented a comparatively small proportion of the British population as a whole. Moreover, in looking at crime and at an alleged decrease in labour force participation, Murray did not draw attention to ethnicity, relating trends instead to social class. He wrote that 'the England in which the family has effectively collapsed does not consist just of blacks, or even the inner-city neighbourhoods of London, Manchester, and Liverpool, but lower working-class communities everywhere'.[39] Frank Field, the maverick Labour MP for Birkenhead, omitted ethnicity in his discussion of an 'underclass'. Rather he drew attention to the emergence of record post-war levels of unemployment, the exclusion of the very poor from rising living standards, widening class differentials and a significant change in the attitudes of those in mainstream society towards those who had failed to 'make it'. These changes, argued Field, had combined to 'produce an underclass that sits uncomfortably below that group which is referred to as living on a low income'.[40] The underclass comprised three groups – the long-term unemployed (especially older workers and school-leavers if they had never had a job), single-parent families and elderly pensioners.

Some sociologists mentioned ethnicity in passing. W.G. Runciman, for instance, argued that if the stereotypical member of the underclass in 1910 had been a loafer – a white, male, casual worker living in rented accommodation – in 1980 it was a single mother from an ethnic minority, living in council housing and entirely dependent on state benefit.[41] However, most other sociologists, who were generally sceptical of the concept of the underclass, or at least agnostic, paid little attention to ethnicity. Duncan Gallie argued that there was very little evidence from the labour market that the disadvantages experienced by ethnic minorities, women or the unemployed were of a type that supported the emergence of an underclass. Rather the idea of an underclass relied on lumping together very different types of labour market disadvantage.[42] Other research, such as that by Lydia Morris and Sarah Irwin on Hartlepool in the North East, was based on areas of high unemployment that did not have large ethnic minority populations.[43] Even research on the Republic of Ireland, where unemployment was very high in the 1980s, concluded that there was little evidence for subcultural characteristics; the underclass framework was redundant and it was sufficient to refer to marginalisation and deprivation.[44]

The underclass debates in the United States in the 1980s

This was the complete opposite of the United States, where much of the underclass debate of the 1980s was predicated on the notion of racial difference. One of the first accounts of the underclass appeared in an article in *Time* magazine in August 1977; much of the content was an account of the minority poor in large-scale cities, accompanied by a series of photographs that featured blacks and Hispanics.[45] The American writer, Nicholas Lemann, was interested in the way that a black underclass culture might be linked to the migration of sharecroppers from the South, and subsequent changes in the composition of ghetto areas. He considered successive waves of migration, from the rural South to the urban North in the 1940s, 1950s and 1960s, and then from the late 1960s, when the black working and middle classes began to migrate out of the Chicago ghettos into the suburbs. He argued that several factors then turned the small underclass from the South into the large separate culture that it became and facilitated a descent into 'social disorganisation'.[46] In *Losing Ground* (1984), Charles Murray related trends in unemployment and in out-of-wedlock births to blacks in particular.[47] It was an analysis taken up by commentators, both those who favoured a behavioural analysis and others who took a more structural view.[48]

Arguably, the most important contributor to the underclass debate in the United States has been William Julius Wilson, Professor of Sociology at the University of Chicago. One of Wilson's points was that following the debate about the Moynihan Report (1965), liberals had left discussion of these issues to the conservatives. Nonetheless, Wilson pointed out that poverty in the United States had become more urban, concentrated and firmly entrenched in large cities, especially the older industrial cities with large and highly segregated black and Hispanic residents. This increase in ghetto poverty was mainly confined to cities in the Northeast and Midwest. Wilson's argument was that historical discrimination and migration to large cities that kept the urban minority population relatively young created a problem of weak 'labour force attachment' among urban blacks. Especially since 1970, this had made them particularly vulnerable to industrial and geographical changes in the economy. These problems were particularly severe in the ghetto neighbourhoods of large cities, because the poorest people lived there, and because the areas had become less diversified. Since 1970, inner-city neighbourhoods had experienced a migration of middle- and working-class families to the suburbs. Combined with the increase in the number of poor caused by rising joblessness, this meant that poverty was more sharply concentrated in these areas. The number of inner-city neighbourhoods with poverty rates above 40 per cent had increased dramatically.[49]

Wilson conceded that by the 1980s there was a large sub-population of low-income families and individuals whose behaviour contrasted with that of the general population. In contrast, and in the years before 1960, inner-city communities had shown signs of social organisation. People had a sense of community, they identified with their neighbourhood, and they adopted norms and sanctions against behaviour they regarded as wrong. Wilson argued that the central problem of the underclass was unemployment that was reinforced by an increasing social isolation in impoverished neighbourhoods. What he called 'weak labour force attachment' was caused by two factors – macro-structural changes in the wider society and economy and the social milieu of individuals.[50] Cultural values emerged from specific circumstances, life chances and class structure. Like other writers in the 1960s, he argued that culture and behaviour were an adaptive response to the circumstances that individuals found themselves in. The transmission of these beliefs was part of what Wilson called 'concentration effects', that is, the effects of living in an impoverished neighbourhood. It followed, then, that the problems of the underclass could be most meaningfully addressed by a comprehensive programme that combined employment and social welfare policies, and featured universal rather than race or group-specific measures. Wilson argued that to ignore the term 'underclass' in favour of more neutral terms such as 'working class' was in his opinion to 'fail to address one of the most important social transformations in recent United States history'.[51]

The focus on social exclusion from the mid-1990s

While Wilson had argued that there was a class of people, of whatever ethnic background, who were now behaving differently, as a response to structural changes in the economy, he had no British counterpart. In any case, since the early 1990s, and among academics and policy makers, the term 'underclass' passed out of use, at least in the United Kingdom, and was replaced by the term 'social exclusion'. In December 1997, for instance, in a speech given at Stockwell Park School, in the deprived London borough of Lambeth, the then Prime Minister, Tony Blair, outlined government plans to tackle the problem of social exclusion. The speech marked the launch of the government's Social Exclusion Unit, and the Prime Minister said that 'social exclusion is about income but it is about more. It is about prospects and networks and life-chances. It's a very modern problem, and one that is more harmful to the individual, more damaging to self-esteem, more corrosive for society as a whole, more likely to be passed down from generation to generation, than material poverty'.[52] According to Blair, part of the answer lay in ensuring that those government departments concerned with the development of policy were co-ordinated more effectively. But Blair also argued that it was in people's own interests that social exclusion should be eliminated. The issue was 'as much about self-interest as compassion'.[53]

From its establishment in December 1997, the government's Social Exclusion Unit issued a range of reports on subjects that included truancy and school exclusion, 'rough sleepers', teenage pregnancy and neighbourhood renewal. The thrust of this interpretation was reflected in a plethora of government initiatives that aimed to tackle social exclusion – the Sure Start programme for parents and children, and area-based initiatives such as the Education and Health Action Zones, the New Deal for Communities, the Single Regeneration Budget and many more. Much of the intellectual input to the work of the Social Exclusion Unit was provided by the ESRC Research Centre for Analysis of Social Exclusion (CASE), established at the LSE in October 1997. In this respect, the issue of social exclusion provides a good example of the close ties that New Labour developed with social scientists.

'Social exclusion' was a term that was imported into Britain from France, where exclusion had become the subject of discussion in the 1960s. In 1974, for example, René Lenoir, then Secretary of State for

Social Action in the Chirac Government, estimated that 'the excluded' made up one-tenth of the French population; but it was only in the late 1970s that 'exclusion' was identified as the central problem of the 'new poverty'.[54] Thus the term 'exclusion' referred to the rise in long-term and recurrent unemployment, and also to important changes in social relations – family break-ups, single-member households, social isolation and the decline of traditional class solidarity based on unions, workplaces and networks. In the 1980s, the meanings of 'exclusion', and its counterpart 'insertion', were expanded to cover emerging new social groups and problems, and were increasingly concerned with the integration of ethnic minority groups. The academic, Hilary Silver, describes how young second generation North African migrants from the housing projects of the *banlieues*, the suburbs or outskirts of the city, argued through their various cultural associations that since they lived in France they should have full citizenship rights. An official policy was adopted to integrate migrants; it managed to keep the key elements of republican solidarity discourse, but also tried to marry these with multicultural meanings of integration.[55] Thus, in French public policy, the many meanings of 'exclusion' unfolded in the 1980s.

In the United Kingdom, there was important research by social scientists interested in health, who sought to tease out the relative significance of ethnicity and social class in explaining patterns of health inequalities,[56] but the wider discourse on social exclusion said little about ethnic minorities, in part because of the legacy of the underclass debate. Being a member of an ethnic minority in the United Kingdom was perceived as a risk factor for social exclusion, and ethnic background featured in discussions about educational attainment, but there was little or no echo of the American analysis in which 'the "underclass" was usually taken to consist of several generations of people from ethnic minorities, living in ghettos and in receipt of welfare, cut off from the mainstream of society and representing a threat to it'.[57]

Evaluations of the British policy initiatives found that they had served ethnic minorities poorly. The local programmes and national evaluation of Sure Start, for example, failed to address the question of ethnicity with sufficient rigour or sensitivity. Experiences and practices varied widely, and some local programmes abandoned the attempt to work closely with certain minority groups.[58] Some services needed to be targeted to reach minority groups, and few ethnic minority staff members were employed in senior positions.[59] But despite such findings, initiatives against social exclusion were generally paralleled by those against anti-social behaviour, constructed largely as a 'problem' of the white working-class population. Nor, since May 2010, has ethnicity been highlighted in the Coalition Government's efforts to rehabilitate an estimated 120,000 so-called troubled families.[60]

Conclusion

This chapter has traced the emphasis given, or not given, to ethnicity in changing underclass discourses in the United Kingdom since the early 1970s. In his famous speech, Sir Keith Joseph said nothing about the extent to which families from ethnic minority groups might be caught up in his alleged cycle of deprivation; and ethnicity was also very much neglected by the researchers funded through the Transmitted Deprivation Research Programme. In the 1980s, while underclass stereotypes in the United States were predicated on notions of race, and on evidence of the real disadvantage experienced by ethnic minority groups, the parallel debate in the United Kingdom again said very little about ethnicity. Rather, it focused on the long-term unemployment experienced by sections of the white working-class population. Social exclusion in France was originally concerned with North African migrants, and in the United Kingdom too, ethnicity has attracted attention from researchers, for instance in demonstrating failures of policy initiatives such as Sure Start. However, in other respects, ethnicity was not prominent in the discourse on social exclusion, and the objects of related strands of policy, notably anti-social behaviour, were constructed as a 'problem' of the white community. While there certainly were specific policies for ethnic minority groups in fields such as education, health, housing and employment, these were not much linked with a more general underclass discourse. That again seems to be a difference between the United Kingdom and the United States, where the culture of poverty idea certainly did influence the 'War on Poverty' in the 1960s.

The United Kingdom was comparatively late in witnessing the emergence of a significant ethnic minority population, and it grew but slowly – only 3 per cent of the total population in 1976. That alone renders the experience of the United Kingdom radically different to that of the United States. But leaving demographic differences aside, something else seems to have been going on. In explaining the absence of ethnicity from the discourse in the United Kingdom, cognisance has to be taken of the wider intellectual and ideological context for public policy. The Research Programme on Transmitted Deprivation is interesting because it provides insights into the outlook of a generation of social scientists. There were marked similarities in the approach that many of these researchers took. First, they had a deep-seated hostility to individual, cultural, or behavioural explanations of poverty. Many had taken the Diploma in Social Administration at the LSE in the 1960s and had been inspired by their teachers – Richard Titmuss, Peter Townsend, David Donnison and others – into a lifelong commitment to social justice. Second, and related to that, they had a preference for structural causes, answers and solutions. Third, there was an unwillingness to single out particular social groups as deserving of special, targeted attention. And fourth, because of the notoriety of the culture of poverty theory and the Moynihan Report of 1965, there was scepticism about poverty models imported from the United States.

The implications for ethnicity were that later researchers followed the lead of early researchers, such as John Rex and others in Birmingham, in dissociating themselves from the underclass thesis, arguing that despite the work of William Julius Wilson in the United States, the term was part of a racist discourse, and a vocabulary of coded panic. The legacy of the 'Titmuss paradigm' is clearly important in explaining the stance that social scientists in the United Kingdom took on ethnicity, and the way that this differed from other countries, notably the United States. The reasons for the filtering out of ethnicity from the discourse, which on the face of it may seem counter-intuitive, are partly demographic and empirical, but largely political and ideological.

Notes

1. See, for example, J. Welshman and A. Bashford, 'Tuberculosis, Migration, and Medical Examination: Lessons from History', *Journal of Epidemiology and Community Health*, 60 (2006), 282–4; J. Welshman, 'Compulsion, Localism, and Pragmatism: The Micro-Politics of Tuberculosis Screening in the United Kingdom, 1950–1965', *Social History of Medicine*, 16:2 (2006), 295–312; J. Welshman, 'Importation, Deprivation, and Susceptibility: Tuberculosis Narratives in Postwar Britain', in F. Condrau and M. Worboys (eds), *Tuberculosis Then and Now: Perspectives on the History of an Infectious Disease* (Montreal: McGill-Queen's University Press, 2010), 123–47.
2. See, for example, J. Welshman, *Underclass: A History of the Excluded, 1880–2000* (London: Continuum, 2006).
3. H. Silver, 'Culture, Politics and National Discourses of the new Urban Poverty', in E. Mingione (ed.), *Urban Poverty and the Underclass: A Reader* (Oxford: Blackwell, 1996), 105–38.
4. K. Mann, 'Watching the Defectives: Observers of the Underclass in the USA, Britain and Australia', *Critical Social Policy*, 14:2 (1994), 79–99.
5. J. Macnicol, 'In Pursuit of the Underclass', *Journal of Social Policy*, 16:3 (1987), 293–318.

6. On the First and Second World Wars, see G. Stedman Jones, *Outcast London: A Study in the Relationship Between Classes in Victorian Society* (Oxford: Oxford University Press, Peregrine Books edn, 1971, 1984 edn); Women's Group on Public Welfare, *Our Towns: A Close Up* (London: Oxford University Press, 1943).
7. D.P. Dolowitz, 'Policy Transfer: A New Framework of Policy Analysis', in D.P. Dolowitz with R. Hulme, M. Nellis and F. O'Neill, *Policy Transfer and British Social Policy: Learning From the USA?* (Buckingham: Open University Press, 2000), 9–37.
8. C. Booth, 'The Inhabitants of Tower Hamlets (School Board Division), Their Condition and Occupations', *Journal of the Royal Statistical Society*, L, 2 (1887), 326–401, 365–9.
9. Though see also C. Cox, H. Marland and S. York, 'Emaciated, Exhausted and Excited: The Bodies and Minds of the Irish in Nineteenth-Century Lancashire', *Journal of Social History*, 46:2 (2012), 1–26.
10. P. Starkey, 'The Medical Officer of Health, the Social Worker, and the Problem Family, 1943 to 1968: The Case of Family Service Units', *Social History of Medicine*, 11:3 (1998), 421–41, 440.
11. J. Welshman, 'In Search of the "Problem Family": Public Health and Social Work in England and Wales, 1940–70', *Social History of Medicine*, 9:3 (1996), 448–65.
12. See, for example, J. Welshman, *From Transmitted Deprivation to Social Exclusion: Policy, Poverty, and Parenting* (Bristol: Policy Press, 2007, paperback edn 2012).
13. Department of Health and Social Security (29 June 1972), 'The Cycle of Deprivation' (typescript), 4, para 15.
14. Ibid., 4, para 16.
15. Ibid., 5, para 17.
16. Ibid., 11–17, paras 34–53.
17. M. Rutter and N. Madge, *Cycles of Disadvantage: A Review of Research* (London: Heinemann, 1976), 1–13.
18. Ibid., 257–301.
19. Ibid., 257.
20. Ibid., 284.
21. Social Science Research Council-Department of Health and Social Security, *Transmitted Deprivation: First Report of the DHSS-SSRC Joint Working Party on Transmitted Deprivation* (London: SSRC, 1974), para 18.3.
22. Social Science Research Council-Department of Health and Social Security, *Transmitted Deprivation: Third Report of the DHSS-SSRC Joint Working Party on Transmitted Deprivation* (London: SSRC, 1977), 11, para 5.4.
23. A. Little and D. Robbins, *'Loading the Law': A Study of Transmitted Deprivation, Ethnic Minorities and Affirmative Action* (London: Commission for Racial Equality, 1982), 8–9, 11–19.
24. A. Little and D. Robbins, 'Racial Disadvantage: Transmission and Counteraction', in M. Brown (ed.), *The Structure of Disadvantage* (London: Heinemann, 1983), 72–100.
25. M. Brown and N. Madge, *Despite the Welfare State: A Report on the SSRC/DHSS Programme of Research into Transmitted Deprivation* (London: Heinemann, 1982), 58–60, 86–9, 99–101, 112–13, 134–5, 190, 216–17, 224–5.

26. D. Donnison, *The Politics of Poverty* (Oxford: Martin Robertson, 1982), 19–21.
27. Little and Robbins, 'Racial Disadvantage: Transmission and Counteraction', 74.
28. O. Lewis, *La Vida: A Puerto Rican Family in the Culture of Poverty – San Juan & New York* (London: Secker & Warburg, 1966, 1967 edn), xxxix.
29. Ibid., xli.
30. Ibid.
31. F.E. Frazier, *The Negro Family in the United States* (Chicago, IL: University of Chicago Press, 1939, rev. and abridged edn 1948, 1966 edn).
32. D.P. Moynihan, *The Negro Family: The Case for National Action* (Washington, DC: US Department of Labor, 1965).
33. Ibid.
34. See, for instance, W.J. Wilson, *The Truly Disadvantaged: The Inner City, the Underclass, and Public Policy* (Chicago, IL: University of Chicago Press, 1987), 3–10, 12; A. Deacon, '"Levelling the Playing Field, Activating the Players": New Labour and the "Cycle of Disadvantage"', *Policy and Politics*, 31:2 (2003), 123–37.
35. Rutter and Madge, *Cycles of Disadvantage*, 30.
36. A. Giddens, *The Class Structure of the Advanced Societies* (London: Hutchinson, 1973), 112, 184.
37. J. Westergaard and H. Resler, *Class in a Capitalist Society: A Study of Contemporary Britain* (London: Heinemann, 1975), 356.
38. J. Rex and S. Tomlinson, *Colonial Immigrants in a British City: A Class Analysis* (London: Routledge & Kegan Paul, 1979), 16, notes 9, 33, 104.
39. C. Murray, *Underclass: The Crisis Deepens* (London: Institute of Economic Affairs, 1984), 11.
40. F. Field, *Losing Out? The Emergence of Britain's Underclass* (Oxford: Basil Blackwell, 1989), 2.
41. W.G. Runciman, 'How Many Classes Are There in Contemporary British Society?', *Sociology*, 24 (1990), 388.
42. D. Gallie, 'Employment, Unemployment, and Social Stratification', in D. Gallie (ed.), *Employment in Britain* (Oxford: Blackwell, 1988), 467–74, 488.
43. L. Morris and S. Irwin, 'Employment Histories and the Concept of the Underclass', *Sociology*, 26 (1992), 401–20.
44. B. Nolan and C.T. Whelan, *Resources, Deprivation and Poverty* (Oxford: Oxford University Press, 1996), 152–78.
45. Anon., 'The American Underclass: Destitute and Desperate in the Land of Plenty', *Time*, 110 (29 August 1977), 34–41.
46. N. Lemann, 'The Origins of the Underclass: Part 1', *Atlantic Monthly* (July 1986), 35, 53.
47. C. Murray, *Losing Ground: American Social Policy, 1950–1980* (New York: Basic Books, 1984).
48. See, for instance, C. Cottingham, 'Introduction', in C. Cottingham (ed.), *Race, Poverty and the Urban Underclass* (Lexington, KY: Lexington Books, 1982), 3; K.B. Clark and R.P. Nathan, 'The Urban Underclass', in National Research Council, *Critical Issues for National Urban Policy: A Reconnaissance and Agenda for Further Study* (Washington, DC: National Research Council, Committee on National Urban Policy, 1982), 33; J.D. Kasarda, 'Structural Factors Affecting the Location and Timing of Urban Underclass Growth',

Urban Geography, 11 (1990), 234–64; D.G. Glasgow, *The Black Underclass: Poverty, Unemployment, and Entrapment of Ghetto Youth* (San Francisco, CA: Jossey-Bass, 1980), vii; A. Pinkney, *The Myth of Black Progress* (Cambridge: Cambridge University Press, 1984), 115–34.

49. Wilson, *The Truly Disadvantaged*, 163, 165–87.
50. Ibid., 18–19, 158–9.
51. Ibid., 7.
52. Welshman, *Underclass*, 183.
53. Ibid.
54. H. Silver, 'Social Exclusion and Social Solidarity: Three Paradigms', *International Labour Review*, 133: 5–6 (1994), 531–5.
55. Ibid.
56. See, for instance, R. Williams, W. Wright and K. Hunt, 'Social Class and Health: The Puzzling Counter-Example of British South Asians', *Social Science and Medicine*, 47:9 (1998), 1277–88.
57. T. Burchardt, J. Le Grand, and D. Piachaud, 'Introduction', in J. Hills, J. Le Grand, and D. Piachaud (eds.), *Understanding Social Exclusion* (Oxford: Oxford University Press, 2002), 2.
58. G. Craig with S. Adamson, N. Ali, S. Ali, A. Dadze-Arthur, C. Elliott, S. McNamee and B. Murtuja, *Sure Start and Black and Minority Ethnic Populations* (London: DfES, 2007), 1.
59. L. Ward, 'Sure Start Failing Ethnic Minorities, Says Report', *Guardian*, 10 July 2007, 9.
60. http://www.number10.gov.uk/news/troubled-families-speech/ [accessed 5 February 2013].

Index

Note: Page references with letter 'n' followed by locators denote note numbers.

195